DISCARDED

Springer Series in Physical Environment

2

Managing Editor
D. Barsch, Heidelberg

Editors
I. Douglas, Manchester · F. Joly, Paris
M. Marcus, Tempe · B. Messerli, Bern

Advisory Board
F. Ahnert, Aachen · V. R. Baker, Tucson
R. G. Barry, Boulder · H. Bremer, Köln
D. Brunsden, London · R. U. Cooke, London · R. Coque, Paris
Y. Dewolf, Paris · P. Fogelberg, Helsinki · O. Fränzle, Kiel
I. Gams, Ljubljana · A. Godard, Meudon · A. Guilcher, Brest
H. Hagedorn, Würzburg · J. Ives, Boulder
S. Kozarski, Poznań · H. Leser, Basel · J. R. Mather, Newark
J. Nicod, Aix-en-Provence · A. R. Orme, Los Angeles
G. Østrem, Oslo · T. L. Péwé, Tempe · P. Rognon, Paris
H. Rohdenburg†, Braunschweig · A. Semmel, Frankfurt/Main
G. Stäblein, Bremen · H. Svensson, København
M. M. Sweeting, Oxford · R. W. Young, Wollongong

Ognjen Bonacci

Karst Hydrology
With Special Reference to the Dinaric Karst

With 119 Figures

Springer-Verlag
Berlin Heidelberg New York
London Paris Tokyo

Professor Dr. OGNJEN BONACCI
Civil Engineering Institute
Faculty of Civil Engineering Sciences
University of Split
58000 Split, V. Masleše b.b.
Yugoslavia

Translated by
Zjena Vidović-Čulić

ISBN 3-540-18105-9 Springer-Verlag Berlin Heidelberg New York
ISBN 0-387-18105-9 Springer-Verlag New York Berlin Heidelberg

Library of Congress Cataloging-in-Publication Data. Bonacci, Ognjen, 1942– Karst hydrology, with special reference to the Dinaric karst. (Springer series in physical environment ; 2) Bibliography: p. Includes index. 1. Hydrology, Karst. 2. Hydrology, Karst—Dinaric Alps. I. Title. II. Series. GB843.B66 1987 551.49

This work is subject to copyright. All rights are reserved, whether the whole or part of the material is concerned, specifically the rights of translation, reprinting, reuse of illustrations, recitation, broadcasting, reproduction on microfilms or in other ways, and storage in data banks. Duplication of this publication or parts thereof is only permitted under the provisions of the German Copyright Law of September 9, 1965, in its version of June 24, 1985, and a copyright fee must always be paid. Violations fall under the prosecution act of the German Copyright Law.

© Springer-Verlag Berlin Heidelberg 1987
Printed in Germany

The use of registered names, trademarks, etc. in this publication does not imply, even in the absence of a specific statement, that such names are exempt from the relevant protective laws and regulations and therefore free for general use.

Typesetting: K + V Fotosatz GmbH, Beerfelden.
Offsetprinting and Bookbinding: Konrad Triltsch, Graphischer Betrieb, Würzburg.
2132/3130-543210

To my family

Preface

Karst is characterized particularly by special landforms and sub-surface drainage. The various actions of water result in numerous variations of surface and sub-surface karst forms. They also bring about distinctive geologic-morphologic forms, and more strikingly, specific flora and fauna. The scientific discipline of hydrology, although a long-established science, cannot easily be applied to karst regions with their very complex drainage system. A special approach is therefore necessary to understand and predict water circulation in these areas.

This is the viewpoint we must adopt if hydrology is to solve the complex problems of karst phenomena. This book can be seen as the appeal of a hydrologist to experts from different scientific disciplines (geology, hydrology, geomorphology, geography, geophysics, meteorology, ecology, civil engineering, forestry, agriculture, etc.) to collaborate towards a better understanding of karst areas.

Evidently, karst phenomena have not been sufficiently and carefully studied worldwide. It is equally true that the first theories on water circulation in karst were developed according to experiences in the Dinaric karst. This can easily be explained. Inhabitants in those areas had no place to which to escape, as was the case in other countries.

The greatest part of this book (ca. 70%) gives a general approach to this phenomenon of karst; however, numerous practical examples are also presented. The majority of the examples refer to the Dinaric karst areas, but there are numerous examples from the other parts of the world also (United Kingdom, France, the Soviet Union, Switzerland, etc.). The approach is primarily hydrological, but the book is written so as to build a bridge over the gaps between experts from various scientific fields related to karst. Close scientific cooperation between experts from various countries will ensure an optimal exploitation of water and the conservation of this natural resource for future generations.

The careful and extensive study of the water circulation in karst involves numerous, well-organized measurements and detailed scientific analyses. Man's work in karst regimes has not always been beneficial; on the contrary, it has caused serious damage. This book may help to avoid and minimize these harmful effects.

It covers 15 years of my work, with special emphasis on the research carried out during the last 10 years (1976–1986), i.e. since I came to live in Split and work at the Faculty of Civil Engineering. However, it also includes experiences of experts from Yugoslavia and other countries. Accordingly, I should like to thank them all for sharing these experiences and their knowledge, and apologize to any colleagues I have inadvertently not mentioned. My special thanks to my close collaborators in writing this book: Sanja Roglić-Perica for the many numerical and computer calculations, Zjena Vidović-Čulić for her translation into English, Myrna Sviča-rević for the proof-reading of the English text, Suzana Blažević for typing and Zlata Rogošić for the well-drawn figures.

This book would never have been written without the support of my family: my wife Tanja and my children Duje, Dunja, and Lada. They have been most affected by the problems related to the writing of this book, but I hope they will also benefit from it. I should like to thank Professor Dietrich Barsch (Heidelberg), Professor Jean Nicod (Aix-en-Provence), and Dr. Marjorie Sweeting (Oxford) for helpful suggestions which contributed to increasing the quality of the book. Finally, I should like to express my deep gratitude to the publisher, Springer-Verlag, and particularly to Dr. W. Engel for his beneficial cooperation in the final formation of this book. The publishers and Dr. Engel immediately and clearly grasped the importance of the specific approach presented in this book and its significance in the field of physical environment.

Split, August 1987					OGNJEN BONACCI

Contents

1	**Introduction**	1
2	**Karst Terminology – Definitions**	4
3	**Principles of Karst Groundwater Circulation**	18
3.1	Subcutaneous Zone (Epikarstic Zone)	28
3.2	Karstification Depth and Karst Capacity for Water Storage	36
4	**Karst Springs**	49
4.1	General Concept and Classification	49
4.2	Discharge Curves	67
4.3	Hydrograph Analysis	75
4.4	Determination of the Catchment Area	81
4.5	The Origin of Brackish Karst Springs	97
5	**Swallow Holes (Ponors)**	103
5.1	Introduction	103
5.2	Determination of the Swallow Capacity of Ponors	103
6	**Natural Streamflows in Karst**	116
6.1	Interaction Between Groundwater and Water in the Open Streamflows	116
6.2	Hydrologic Regime of Rivers in Karst	120
6.3	Water Losses Along the Open Streamflows in Karst	124
7	**Hydrologic Budget for the Poljes in Karst**	136
8	**Water Temperature in Karst**	141
8.1	Introduction	141
8.2	Groundwater Temperature in Karst	141
8.3	The Water Temperature of Springs and Open Streamflows in Karst	145

9	**Man's Influence on the Water Regime in the Karst Terrains**	150
9.1	Water Storage	151
9.2	Increase in the Capacity of the Outlet Structures	159
9.3	Surface Hydrotechnical Works	167
9.4	Action on the Groundwater	168
9.5	Usage of the Karst Spring Water	169
9.6	The Development of the Brackish Karst Springs	171

References 175

Geographical and Subject Index 181

1 Introduction

Karst represents a specific area consisting of surface relief and a surface-underground hydrographic network resulting from the water circulation and its aggressive chemical and physical action in joints, fractures and cracks along the layers of soluble rocks, such as limestone, chalk and dolomite as well as gypsum and salt. Karst is characterized by soluble rocks located near or at the surface. The karstification process results from the physical and chemical water action on the solution and transportation of elements from the rocks. The power of water solubility in contact with rocks depends on the water temperature and its chemical composition, with the dominant component being CO_2. The coarseness of grain size is an important factor in limestone dissolution. It effects the chemical quality of karst waters (Sweeting 1973). The fineness of grain affects the solubility of the rock. Sweeting (1973) has concluded, after laboratory experiments on dolomites from Ontario, that the finer grained dolomites are about twice as soluble as the coarser grained beds. Owing to specific geologic and geomorphologic, and particularly hydraulic characteristics, karst areas exhibit a specific water circulation which should be studied by appropriate methods. Karst hydrology is a relatively new scientific discipline if it is considered as an independent scientific branch. It has been so far included in the investigations on the process of water circulation in karst as a part of geology and hydrogeology, but it is developing as an independent discipline. Its present and future development are in close connection with the advances in other disciplines primarily geology, hydrogeology, hydraulics, geophysics, chemistry, hydrometry, climatology and statistics.

Karstified rocks can be found in all parts of the world. In certain regions they are quite frequent and cover wide and deep areas (e.g. in Yugoslavia along the Adriatic Sea), whereas in other regions they are rare and appear only in certain areas, most often as shallow surface karst (countries of northern Europe and South America). Karst is estimated to cover 20 to 25% of the surface of all continents. Milanović (1981) describes in detail the most significant areas covered by karst in the world.

It can be stated, without exaggeration, that the karst areas of Yugoslavia represent a cradle of a systematic study of karst in the world, considering the water circulation in the cracked and fissured areas. The first theoretical and pratical engineering investigations were carried out either in the Yugoslav karst or by the Yugoslav researchers, Cvijić (1893), Gavazzi (1904), Grund (1903) and Waagen (1910). This seems quite natural if we bear in mind the fact that ca. 75 000 km^2 of the surface in Yugoslavia (i.e. 35% of the total area) are strongly influenced by karstification processes. Figure 1.1 represents a schematic situation of the parts of Yugoslavia covered by karst. The Dinaric karst, situated along the coast by the

Fig. 1.1. The karst areas in Yugoslavia

Adriatic Sea is particulary well developed and deep with very specific karst geomorphologic forms. In the past the necessity of investigating the water circulation in karst in Yugoslavia has been caused by the fact that men have continually inhabited these areas. While in other parts of the world the karst regions are mostly unpopulated, in Yugoslavia some karst areas are densely populated. The needs of people living in those areas have brought about the development of theoretical and practical solutions in karst hydrology. With the increase in the number of inhabitants on the surface of Earth, the karst areas will become more densely populated, and this fact entails the settlement of people in the karst areas. Thus, the interest in the study of karst hydrology will increase, especially with regard to water circulation, its phenomena and quantity.

The water circulation processes in karst have been studied by numerous researchers; however, it can be noted that, in general, the hydrologic aspect of this problem has not been dealt with carefully enough so far, and hydrologic methodology has not been appropriately applied in practice. Consequently, this book stresses the hydrologic aspect of the problem, i.e. an attempt to define quantitatively the relations and quantities of the surface and groundwater in the karst medium. It should be borne in mind, however, that karst phenomena, and the water circulation in it can be studied in detail only by applying an interdisciplinary approach. Karst is a very heterogeneous space medium which can be explained only be applying a great number of measurement points and by monitoring and measuring various phenomena: the water levels, water discharge and velocity, chemical composition of water, its temperature, concentration of suspended particles and the microorganisms in water. Consequently, karst should be distinguished from the hydrologically more homogeneous non-karst areas which can be treated using physical models or at least as a "grey box". Karst still calls for a system approach to an analysis performed by observing the water inflow into and out of the system and by treating the water circulation in the system as a "black box".

2 Karst Terminology – Definitions

The primary objective of this book is to study the hydrologic aspect of the problems of water circulation in the karst. Therefore, karst phenomena, and types and characteristics of karst will not be dealt with from the geologic and geomorphologic point of view. The following books or papers refer to the above problems: Zötl (1974), Milanović (1981), Bögli (1980), LeGrand and Stringfield (1973), Sweeting (1973) and LeGrand (1983). Figure 2.1 is taken from the latter two papers and represents six general topographic categories which can be developed in the karst regions. The authors believe that these typical karst forms include all phenomena and are valid for the whole world. This division, as any other, has its advantages and disadvantages, but they are not important from the point of view of hydrology. The well-developed fissures of the karst areas, and consequently a very fast and great water sinking strongly affect the distinctive hydrologic behaviour of water in the karst terrains to be distinguished from the circulation in porous media. The possibility of surface flow is either unlikely or completely eliminated. In these considerations we should take into account the fact that apart from the bare karst (appearing in southern Yugoslavia, Dalmatia and Herzegovina) there are areas of covered karst (Slovenia and Gorski Kotar) with rich vegetation and a layer of covering soil up to 1 m deep, in some places even several meters. In such situations the sinking is significantly slower, whereas all other processes of water circulation are identical to those in the bare karst.

The karstification processes are caused by the contact of water with the soluble and porous rocks. The more intensive tectonic processes condition the development of stronger karstification and a more intensive water circulation. The essential prerequisite for the formation of karst forms is the temperature and abundant precipitation. The other factors more significantly affecting the karstification process are the chemical composition of water, the velocity of water circulation and the climate in general. The karstification processes are more intensive if there is a thick layer of humus above the carbonate rocks. This is explained by the fact that this thicker layer of humus favourably affects the production and accumulation of CO_2, which in the phase of a chemical process participates in the solution of carbonate rocks. Generally speaking, karst terrains are dry and unfavourable for the living conditions of human beings, but there are some areas with abundant precipitation, ranging from 600 mm to 3000 or even 4000 mm a year (Crna Gora, southwest of Yugoslavia). This seems illogical at first sight, but it can be explained by excessively fast infiltration and by considerable water percolation, i.e. very fast vertical sinking of water from the surface, through the aeration zone to the underground water level.

Karst Terminology – Definitions

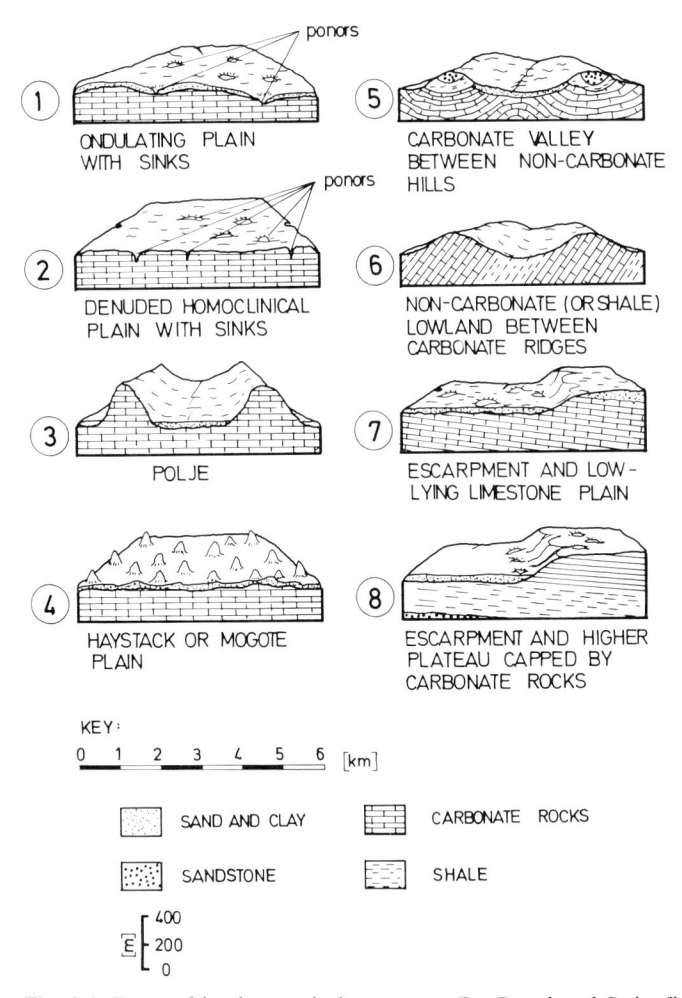

Fig. 2.1. Types of landscapes in karst areas. (Le Grand and Stringfield 1973)

The primary, textural or intergranular porosity of carbonate rocks has formed open spaces in rocks created during the period of their genesis, sedimentation and petrification. The term secondary porosity is used for the spaces created in carbonate rocks in the post-petrification period. This phenomenon includes pore spaces created by the solution processes or by some other ways (i.e. fractures). The number and dimensions of pores increase gradually with time, particulary due to the solubility of rocks and the physical action of water. Hence, the porosity of carbonate rocks per unit of area increases in time. The size and number of pores, and the shape of joints, are of primary importance for the definition of the rock porosity. The existence of voids in the rocks implies their porosity. The water can flow through the pores if they are connected and if their dimensions are such to enable the water to overcome the adhesion of molecular forces. The fact that carbonate rocks are permeable is more important than their porosity for the hydrologic approach to karst investigations.

Further on we will define several basic terms pertaining to the relation between water and carbonate rocks. The definitions are taken over from Castany, from the UNESCO (1984) publication.

Gravitational water is the water, which due to gravity, penetrates and flows into the underground system. Its volume is designated by V_g.

Retained water is the part of the underground water in karst pores retained there by the surface tension or molecular attraction; its volume is expressed as V_r. The relation between the volume of the gravitational water V_g and the retained water V_r is the function of the pore size and the mineralogic characteristics of the rocks. If the pores are smaller than 4 to 5 µ, the gravitational movement of water is practically blocked.

Specific retention represents the relation between the volume of the retained V_r and the total volume of the considered rock mass V.

Total porosity, n, is the relation between the volume of all pores V_v/V is expressed in percentages. In water-saturated rocks $V_v = V_g + V_r$, i.e. it represents the sum of gravitational and retained water. The amount of the total porosity of carbonate rocks varies considerably and ranges widely from 0.2 to 45%.

Effective porosity, n_e, applies only to connected voids with water circulation. It is expressed in percentages of the total volume V in the following way:

$$n_e = (V_g/V)\ 100. \tag{2.1}$$

Effective porosity changes in time. If we consider an idealized case of exclusively underground water inflow into the carbonate mass, the change in effective porosity in time can be presented as in Figure 2.2. The effective porosity is an index of connection between the voids which can in time be filled by gravitational water.

Storage coefficient, S, is the volume of water released from the prism of a unit cross-sectional area whose height is the total thickness of the aquifer due to change in the piezometric level for 1 m. In a confined aquifer the storage coefficient ranges from 10^{-3} to $10^{-5}\ m^3$, whereas in a free water surface aquifer the storage coefficient corresponds to the volume of the gravitational water in 1 m³ of the aquifer. The graphic representation of the storage coefficient in the free and confined aquifer is given in Figure 2.3 A, B (Castany – UNESCO 1984).

Specific yield is defined as the change in the water quantity in the storage per unit of area of the free surface of the aquifer. It is the result of the unit change in the height of the groundwater level. The specific yield and storage coefficient increase with time.

The storage capacity of karst carbonate rocks can be determined on samples in the laboratory or by field investigations. It represents the estimation of the water quantity which can be tapped from one part of the aquifer. In wide aquifers with a free water surface it approximately represents the volume of the free water. Since, from the hydrologic point of view, the term effective porosity n_e is very interesting and directly connected with the storage capacity and the general capacity of karst rocks for the water storage, it will be dealt with in detail in Section 3.2.

The previously analyzed terms are not related to the dynamic features of water circulation through the karst carbonate rocks, but they exclusively refer to the relationship and quantity of water in the system of fissures. The essential

Karst Terminology – Definitions

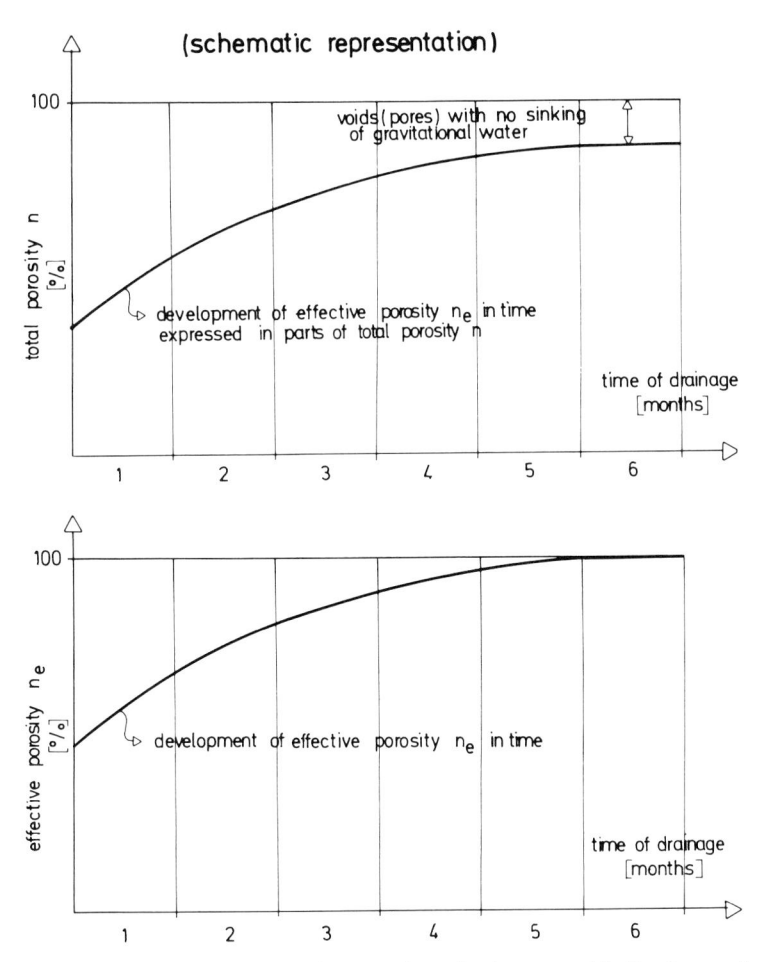

Fig. 2.2. Development of effective porosity n_e in time – an idealized case of permanent inflow into the underground of the carbonate rocks

phenomena related to dynamic characteristics, particularly to the velocity of water circulation, will be described later on.

Permeability, K, is the property of rocks forming an underground reservoir, to allow water to flow with an appreciable velocity due to the existence of the pressure gradient. Permeability depends on the shape and size of the voids. Permeability K has the dimension m s^{-1} and can be defined by the following expression:

$$K = Q/(A \cdot I) , \tag{2.2}$$

where Q denotes the water discharge passing in a second through the cross-sectional area A, expressed in m^2 under a unit hydraulic gradient I, at a temperature of 20 °C.

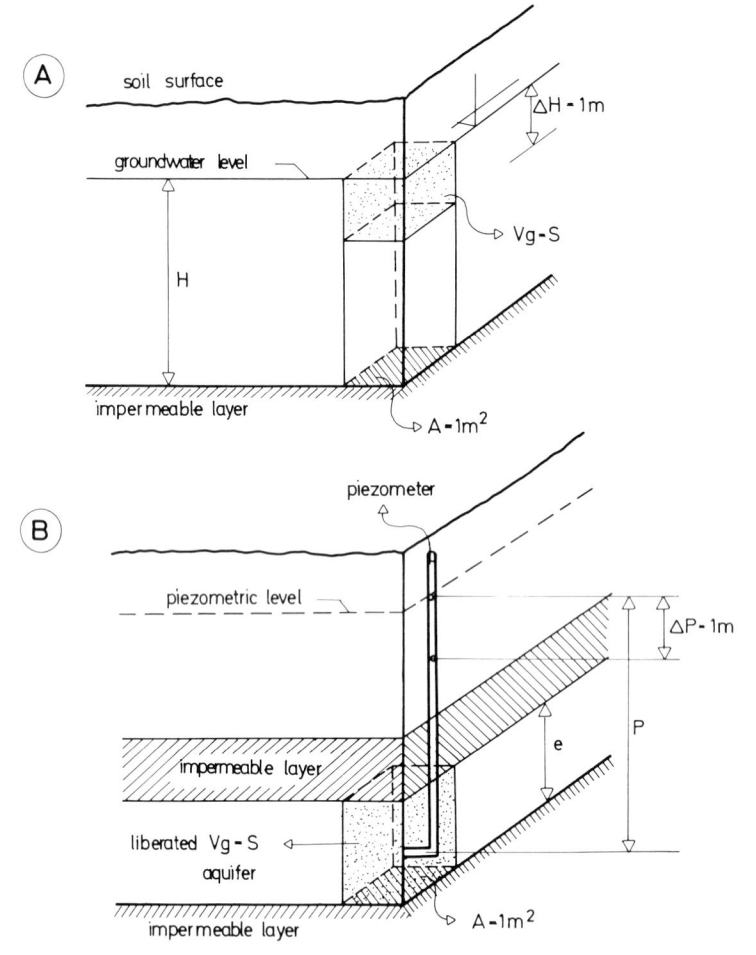

Fig. 2.3 A, B. Storage coefficient in a free surface (**A**) and confined (**B**) aquifer. (Castany-UNESCO 1984)

The coefficient of transmissivity, T, is the water inflow at the prevailing water temperature through the vertical cross-section of the aquifer of one unit width, entirely saturated by water at the hydraulic gradient of 100%. Its dimension is $m^2 s^{-1}$, and it can be used to estimate the aquifer capacity.

The velocity of water circulation in karst primarily depends on whether the water flows through a system of developed karst fissures and channels or through a system of small cracks. The karst porosity is not great, and consequently the water flow should be quite slow. However, well-developed paths create conditions for quick circulation. It is very difficult to define an average velocity of water circulation in karst, primarily as neither the exact paths of the water flow nor their actual length are generally known. Therefore, the so-called distance or fictitious velocity v^* in $m\,s^{-1}$ is defined by the following expression:

$$v^* = L/t , \qquad (2.3)$$

where L expresses the shortest (horizontal) distance between the two points, and t is the travelling time of water between them. This method of determination is most frequently applied in establishing the relations: swallow holes-springs, piezometers-springs or piezometers-piezometers by dyeing tests. According to Knežević (1962), the measured distance velocities in the Dinaric karst are from 0.2 to 20 cm s^{-1}. Milanović (1981) states that these velocities range from 0.002 to 55.2 cm s^{-1}. The average distance velocity from both authors is 6 to 7 cm s^{-1}. Vlahović (1983) presents the measurement data for fictitious velocity obtained by dyeing tests carried out in the ponors in the Slano and Krupac Poljes (the Zeta River Catchment) whose velocities range from 0.89 to 4.58 cm s^{-1}. Magdalenić et al. (1986) give distance velocities from 0.8–2.0 cm s^{-1} measured in Istra at the Bulaž Spring. All the measurements were carried out during low groundwater levels after long periods of droughts. Small distance velocities are one of the basic features of water circulation in the Dinaric karst (Yugoslavia).

The velocity of water circulation in karst significantly varies with the change of the hydrologic conditions of the aquifer. When the aquifer is full, the circulation is much faster than during long-lasting periods of droughts. Since the water velocity in karst is most often measured by dyeing tests, which can determine only distance velocity, the factors influencing the change in velocity are emphasized. Brown and Ford (1971) defined the possible combinations of the flow network between the points of the entrance and exit of the tracers. Figure 2.4A presents five types of networks. Figure 2.4B shows the possibility of the existence of underground storage basins which significantly affect the water flow velocity as well as the shape of the outcoming tracer wave. They essentially reduce the velocity, and lengthen and flatten the outflowing wave. Figure 2.4C presents the case of a connection between the main karst system and one of its subsystems. Such, and even more complex, situations are not rare, and in a way they belong to one of the types of the network II to V, as plotted in Figure 2.4A. It should be noted that depending on the water level in the aquifer of the main system and its subsystems, the circulation may take place in both directions. All the above facts point to the complexity and the possible errors in the determination and interpretation of the direct field measurements of the water flow velocity in karst.

Brucker et al. (1972) measured the velocity of water flowing along the vertical wall of the shafts in West Virginia, Kentucky and Alabama, and obtained the results presented in Figure 2.5. The water flow velocity increased during the wet period, whereas Fig. 2.5 presents the cases of velocity measurements after long dry periods. Gale (1984) analyzed the morphology of dissolution-bed form assemblages and hydraulically-transported sediments found within conduits in carbonate aquifers in North America and Great Britain. These values were used to define numerous hydraulic conditions under which conduit flow occurs.

Mean values of flow velocity have been calculated. Velocities always occur in the turbulent regime, and they range from 0.03 to 1.21 m s^{-1}, whereas discharges vary from 0.13 to 9.14 m^3 s^{-1}. Gale (1984) found that using indirect methods of measurements (bed form-erosional features and hydraulically-transported sediments) it can be generally concluded that velocities in the conduits range from 0.01 to 1 m s^{-1}. The velocities in a diffuse flow, however, belong to a laminar regime. Atkinson (1977) defined the velocity of the diffuse flow in a carbonate

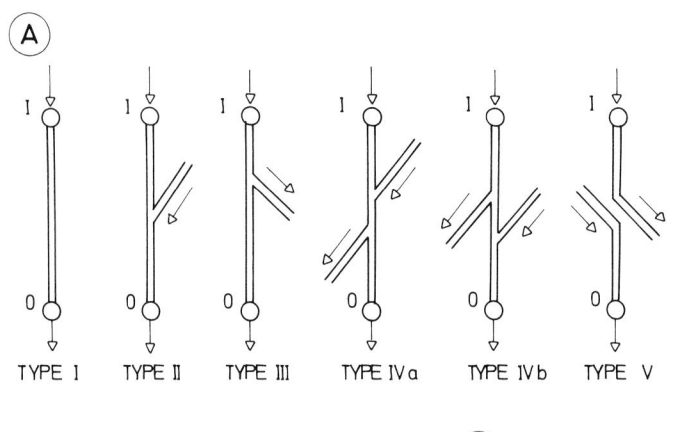

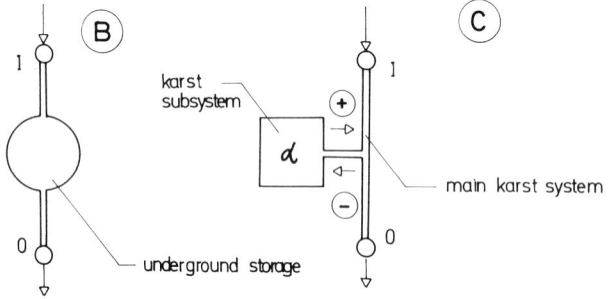

Fig. 2.4 A–C. Possible types of relation between inflow (I) and outflow (O) in the karst system. **A** Types of flow network (Brown and Ford 1973); **B** presence of underground storage; **C** connection between the main system and its subsystem

aquifer of Mendip Hills, UK as 1.03×10^{-3} m s^{-1}, and Foster and Milton (1974) determined the maximum velocity of the diffuse flow as 1.97×10^{-3} measured in the most permeable parts of the Cretaceous Chalk of East Yorkshire, UK.

Fig. 2.6 represents a schematic cross-section of a typical karst system with various surface and underground phenomena. Surface karst forms are: karren, sinkholes (dolines), dry valleys and poljes (Vlahović 1983).

Karren occur in bare carbonate, gypsum and salt karst rocks and are separated by narrow ridges. They also occur under soil cover. They have been formed by an erosive action of water (chemical and physical), and appear mostly in cracked limestone rocks susceptible to the development of such forms, but there are many other types of karren which are independent of cracked limestones. Considering the water flow they represent the first significant paths along which water, passing through the other forms, sinks underground. These are characteristic features for

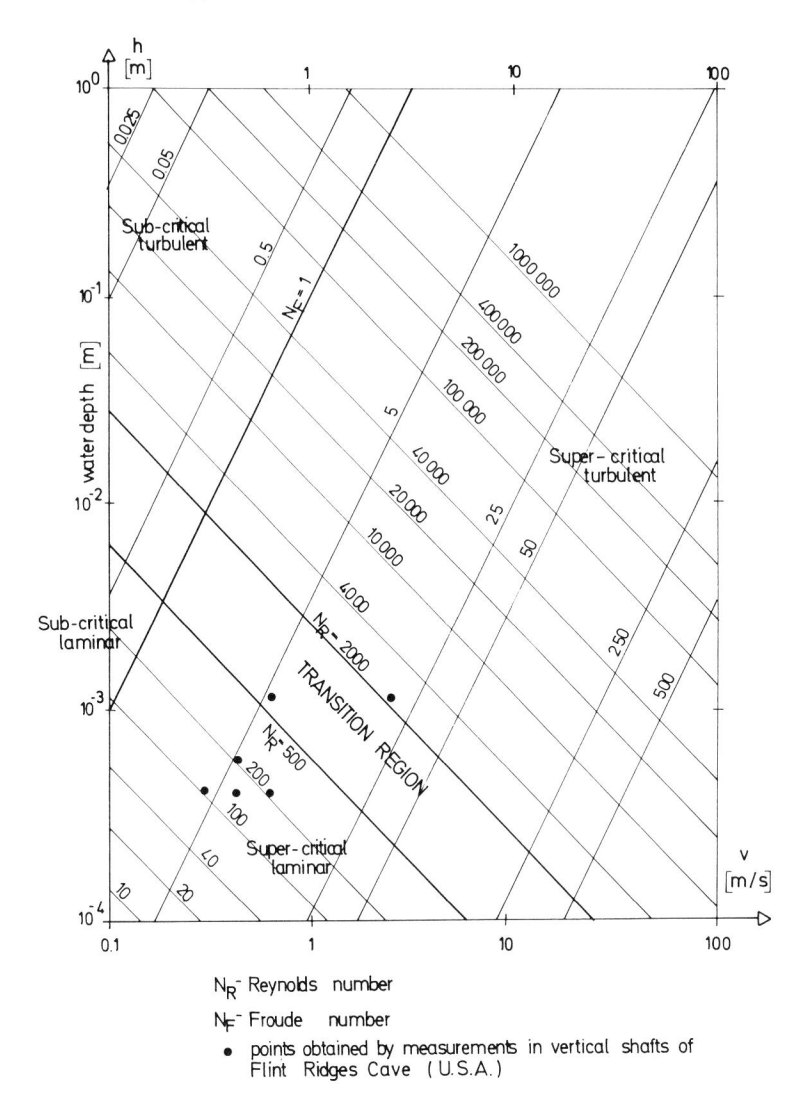

Fig. 2.5. Fence diagram showing flow regimes for wide open channels. (Bruckner et al. 1972)

the hard and bare karst where the surface flow is not actually possible. The interspaces between karren are from several centimeters up to 2 m and more wide, and their depth ranges from 2 m to frequently several meters. Terrains covered by karren are most often situated along the boundaries of the poljes in karst and hardly accessible. In the interspaces there is a layer of soil penetrating deeper into the underground which plays a significant role in the subsequent karstification processes. Karren themselves are the most evident result of the current karst processes and point to the possibility of the appearance of other karst phenomena important for hydrology.

12   Karst Terminology – Definitions

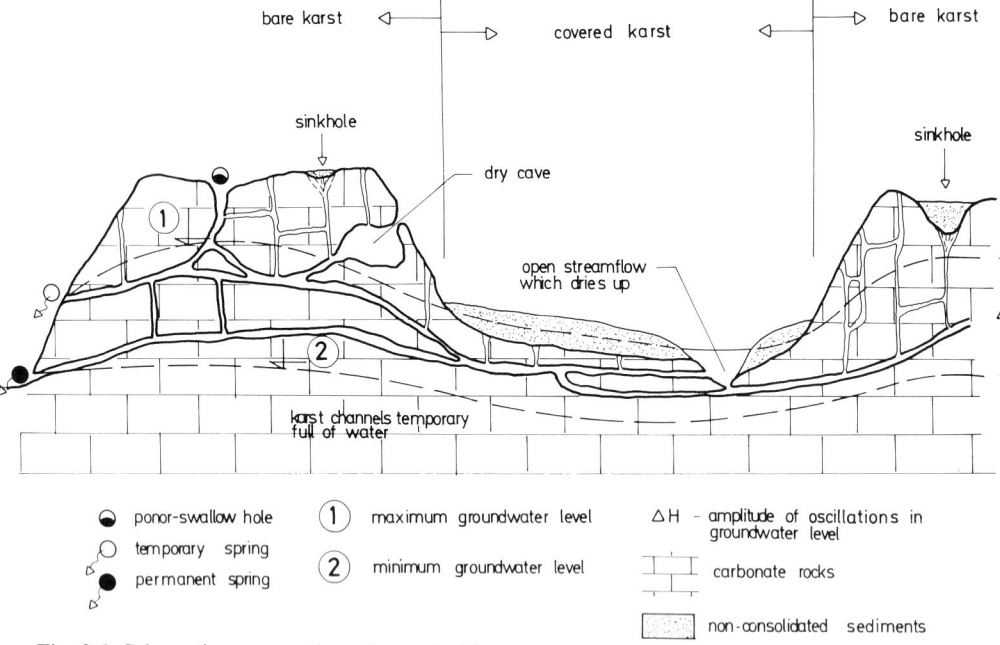

Fig. 2.6. Schematic cross-section of a typical karst system

Dolines (*sinkholes*) are cone-like karst hollow forms with a diameter ranging from a few meters to 100 m, up to 10 m deep. The deepest sinkholes in Yugoslavia are the Blue and Red Lakes near Imotski, about 200 and 500 m deep respectively. Sinkholes are formed apart from the erosive chemical action of water, by collapse of the floors in caverns and karst channels (Milanović 1981). Williams (1983) explains the formation of dolines with regard to the influence of a subcutaneous zone in karst. Figure 3.11 is related to that explanation and the details are presented in the next chapter. Dolines appear mostly in groups, frequently concentrated along the lines of faults. Only a small number of dolines in the Dinaric karst like the Blue and Red Lake have a bottom below the level of the groundwater. In shallow Dinaric karst and also in deep karst of other parts of the world the doline bottoms are frequently situated below the groundwater level, and consequently those dolines are often flooded.

Dry valleys in karst are elongated recesses and valleys, at the bottom of which there are dolines, jamas (shafts) and caves. From the hydrologic standpoint it is interesting to note that there are no permanent watercourses and there are rarely intermittent open streamflows in them. The conditions for the surface flow do not exist since the karstification process has been stronger and faster than the process of the river valley formation. Milanović (1981) gives an example of a typical Trebišnjica dry valley between Hutovo and the Neretva River Valley, where this river used to flow by a surface watercourse. Dry valleys essentially were formed because the groundwater level was low, due to strong karstification processes, and which never, or very rarely, appeared near the surface, thus making the inflow of open streamflows into the underground possible.

Karst Terminology – Definitions

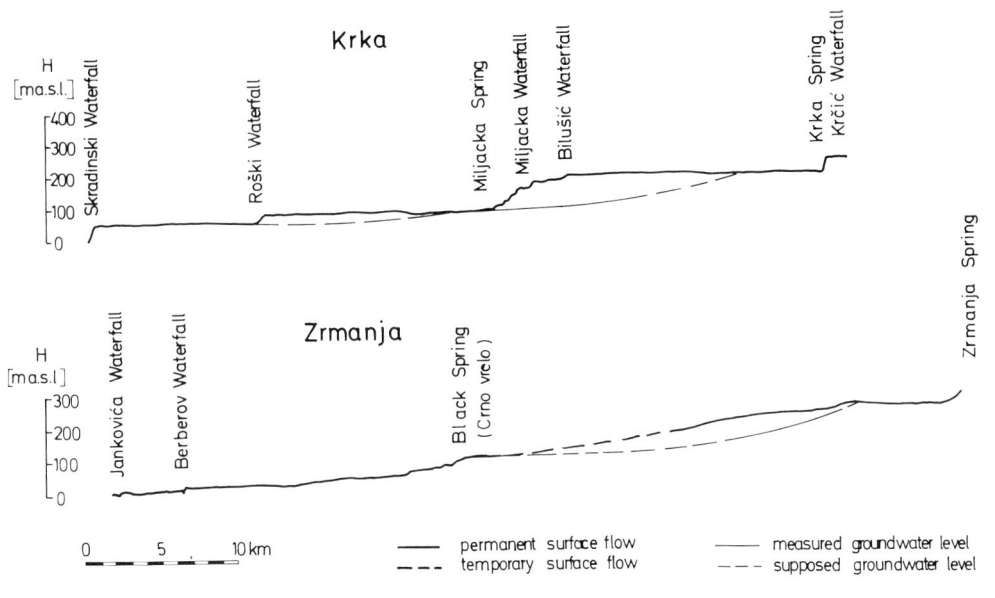

Fig. 2.7. The relation between water level in the open streamflow and the groundwater level along the river courses of the Krka and the Zrmanja Rivers (Yugoslavia). (Fritz and Pavičić 1982)

It should be noted that sometimes in the Dinaric karst permanent watercourses flow beyond the groundwater level, even for 50 m. A characteristic case is the section of the Zrmanja River from Palanka to Ervenik. Similar "suspended" or "perched" stretches can be found on the Krka River from Marasovina to the Miljacka Waterfalls and even further on. While the Zrmanja River dries up on the mentioned section during the dry periods, the Krka never dries up. Furthermore, no significant water losses were established by measurements. The reason why there are practically no losses, and why the losses are fewer and consequently there are no dry valleys, can be explained by the fact that the riverbed of those rivers is filled by their own very fine-grained sediments. These make infiltration more or less impossible, and hence the water losses through the wetted perimeter of the open streamflow are quite rare or do not occur at all. Figure 2.7 presents the results obtained by Fritz and Pavičić (1982) related to the ratio between the water level in the open streamflow of the Krka and Zrmanja and the groundwater level.

Poljes in the karst represent depressions in the limestone karst, generally elliptical with relatively gently sloping bottoms from the spring zone to the swallow-hole zone. They are most often covered by soil belonging to the Neogene and Quarternary sediments i.e. terra rossa. The poljes are also frequently aligned along tectonic and fold axes. Their size varies from the small ones covering 0.5 km^2 to the largest covering 500 km^2. LeGrand (1983) gives the following definition of poljes: "They are flat alluvial valleys bordered by relatively steep bare limestone ridges, ranging from almost a kilometer to several kilometers in width and are somewhat elongated". They have either permanent or temporary springs, often flowing along the longer axis. The surface flow in concentrated streamflows occurs less often along the shorter axis. Poljes in karst can be found in various parts of the world, most often in the Mediterranean countries (Greece, Italy, France, Spain, Morocco, Tunis and Yugoslavia). There is a small number of poljes in Asia, a greater number in Cuba, Jamaica and Canada in the area of Nahanni, whereas there is only one polje (Bögli 1980) in the USA in Tennessee (Grassy Cove). The poljes appear most frequently and have the most specific characteristics in the Dinaric karst of Yugoslavia. This is the main reason why the word polje has passed from Croato-Serbian into all languages and is used as an international technical term in this field. We believe the term "polje in karst" should be used as it seems more appropriate than the formerly accepted term "karst polje". Poljes in the Yugoslav karst (and similarly elswhere) represent the only oases in karst with living conditions favourable for human beings. These regions are covered with arable soil and have either permanent or temporary springs and they are surrounded by bare, rocky, often non-arable terrains. According to Barbalić (1976) the total area covered by closed poljes in Yugoslavia is ca. 1350 km^2 or approximately 2% of the total area covered by karst, estimated to cover about 75000 km^2. Although they are relatively small in size, they are significant from an economic and social standpoint.

From the hydrologic point of view, polje is only a part of a wider system. It cannot, and should not, be treated as a complete system, but only as a subsystem in the process of surface and groundwater flow through the karst massive. Consequently, they cannot be studied properly without establishing the measurement points and devices not only in the polje itself, but also in the karst massive surrounding it, and in the poljes of higher and lower horizons connected with the analyzed subsystem. Poljes in karst are regularly flooded in the cold and wet periods of the year, in Dinaric karst from October to April, and in summer there is not enough water (Bonacci 1985). According to the hydrologic regime, inflows and outflows, poljes can be classified into four basic types: (1) closed poljes; (2) upstream open poljes; (3) downstream open poljes; (4) upstream and downstream open poljes. Flooding is caused by the limited capacity of the outlet structures. Barbalić (1976) states that 35% of the areas covered by polje in Dinaric karst becomes flooded during the year. Figure 2.8 gives a schematic presentation of the mentioned characteristics of poljes in karst.

Underground karst phenomena are jamas (shafts), channels (passages) and caves.

Jamas are shafts of greater dimensions, sinking deep into the carbonate massive. Usually, they can be found within a zone of vertical circulation of water and are filled with air during the greater part of the year. They represent the most important joint systems for the water flow from the surface into the underground (Gospodarič 1976). Very frequently, particularly in the poljes in karst, jamas located in the lower zones function as swallow holes or estavelle. Such jamas are generally circular and are formed by water action. Brucker et al. (1972) explained the characteristics, shape and size of vertical jamas and their role in water drainage in karst. According to these authors, jamas represent channels for the transportation of water efficiently and quickly through the vadose zone. The catchment area of a jama varies significantly. In dry periods it can be quite small,

Karst Terminology – Definitions

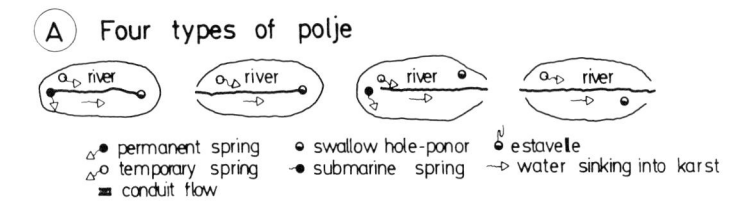

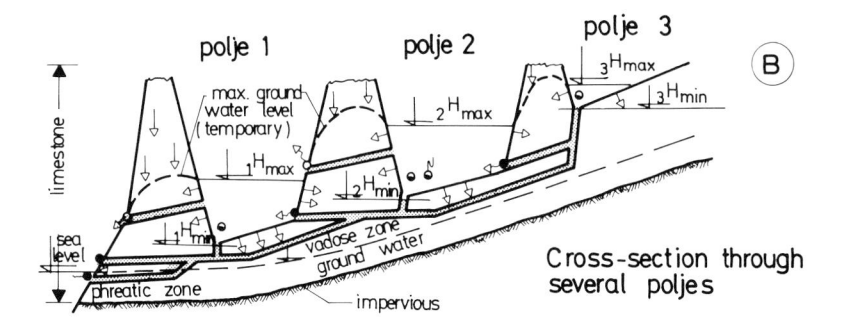

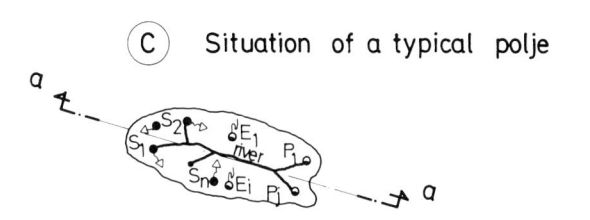

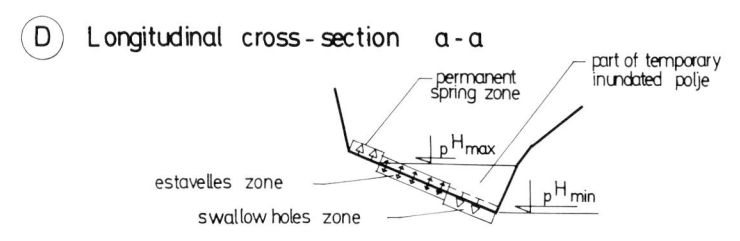

Fig. 2.8 A–D. Schematic presentation of the poljes in the karst. **A** Four types of polje; **B** cross-section through several poljes; **C** situation of a typical polje; **D** longitudinal cross-section *a-a*

even less than 1 ha, whereas during floods, a jama functioning as the main outlet structure can swallow great quantities of water from a large catchment area. Jamas gradually and continually increase in depth as the groundwater level decreases. Essentially they are the forms occurring in a vadose zone, and so far no jamas with a bottom below the lowest groundwater level have been recorded, although some already discovered jamas reach great depth. Milanović (1981) states that the deepest known jama in the entire world is Pierre Saint Martin in

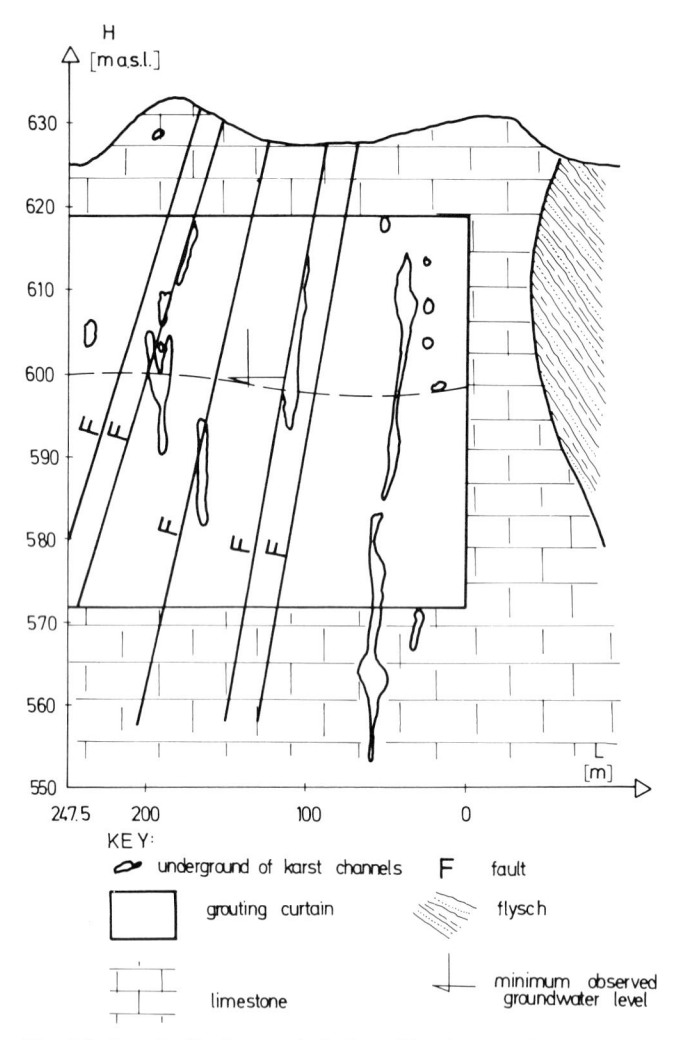

Fig. 2.9. Longitudinal geotechnical profile of a grouting curtain in the zone of the Krupački ponors. (Vlahović 1983)

the Pyrenees (1971 m) between France and Spain, whereas in Yugoslavia the deepest known jama is the Pološka Cave near Tolmin, which is 658 m deep.

Underground karst *channels* (passages) are horizontal or gently sloping pipes of irregular shape and changing dimensions through which water circulates in karst. They very often connect jamas. The dimensions of the karst channels can vary from the order of 1 cm to several meters. Vlahović (1983) presented an unexpectedly great number of underground karst channels in the Slano Polje (Yugoslavia) on the grouting curtain Bročanac – Široka ulica – Orlina. Channels can be found both in the zone beyond the maximum groundwater level and in the zone of groundwater level oscillation, down to a depth of 80 m under the surface. Figure 2.9 presents a longitudinal geotechnical profile of the grouting curtain in the Krupac Polje (Vlahović 1983). Karst erosion is stronger in the zone of intensive water flow in the vertical and horizontal direction than below this zone, which results in a greater number and larger dimensions of channels in that zone. The turbulent flow of water in them ensures a fast inflow of fresh water, while the vegetation phenomenon, closely connected with the presence of water quickens the processes of rotting away which in their turn increase the content of CO_2 in water. Thus, the possibility for chemical erosion of carbonate rocks becomes greater.

Considering the water flow through the karst channels the main factors influencing it are the size and form, as well as the variation of the cross-sectional area perpendicular to the direction of the groundwater flow. Investigations in different parts of the world show great irregularities in the appearance and dimensions of the cross-sections. The observed cross-sections of the channel perpendicular to the direction of the flow should be transformed into ellipses of the same size, in order to analyze their shapes and dimensions in a certain region. Motyka and Wilk (1984) suggest the following classification of the karst channels based on the relation between the length of the horizontal x and vertical y half-axis of the transformed ellipse: 1, vertically elongated $(x/y) < 0.8$; 2, regular $0.8 \leqslant (x/y) \leqslant 1.25$; 3, horizontally elongated $(x/y) > 1.25$. Their investigations carried out in the Triassic carbonates of the Olkusz region (Poland) have shown that there are 40% vertically elongated channels, 23% regular and 37% horizontally elongated channels, whereas the maximum relation between the lengths of the horizontal and vertical half-axis of the ellipse (x/y) ranges from 0.09 to 10.6. Motyka and Wilk (1984) propose the following division of channels considering the cross-sectional area (A in m^2) perpendicular to the direction of the water movement: 1, small $A < 0.25$ m^2; 2, medium $(0.25$ m$^2) \leqslant A \leqslant (0.50$ m$^2)$; 3, large $(0.5$ m$^2) < A \leqslant (1.0$ m$^2)$; 4, very large $A > 1.0$ m^2.

Caves are generally considered as widened karst channels stretching horizontally or gently sloping. Caves provide direct contact of the surface water with the karst underground and vice versa. Caves are quite numerous in Yugoslavia. The most famous for its beauty, dimension, flora and fauna is Postojna, 16670 m long. The largest cave system in the world is Mammoth Cave in the USA in the State of Kentucky. It has not been thoroughly investigated so far, and it is supposed to be of the order of magnitude of 500 km in its total length. The Slana Cave (Vlahović 1983) has strong temporary karst springs which dry up only after long dry periods. Caves do not frequently take over the function of swallow holes.

All the previously mentioned karst phenomena are ideal sites for observing the actual processes of water flow in karst. The devices used are either natural piezometers or measurement devices for continuous or temporary monitoring of various water characteristics, primarily the change of water level. Data gathered on such sites can be exceptionally well used for establishing the process of water flow in a karst system. On the other hand, all the surface, and particularly underground karst phenomena, providing a direct connection between the groundwater and the surface, represent excessively sensitive points for the possible direct and fast inflow of liquid and solid pollutants into the otherwise clear groundwater. This fact should be carefully considered in the study, design and construction of structures on the karst terrains.

3 Principles of Karst Groundwater Circulation

The circulation of water through karstified rocks is similar to water circulation through non-karst terrains under normal hydrologic conditions apart from some specific features. The lateral circulation in karst is effected by the following three concentrating mechanisms (Gunn 1983): (1) overland flow; (2) throughflow, i.e. flow through a layer of soil above limestone; (3) subcutaneous flow. Each of the above flows differs from one situation to another, depending upon the soil cover and vegetation of the catchment, and on the development of the karst processes. The lateral water flow precedes the infiltration and percolation phases, i.e. the vertical sinking of water through the vadose zone down to the groundwater level.

The input mechanisms of the process of groundwater transmission through the vadose zone can be divided into three mechanisms (Gunn 1983): (1) shaft flow of flow through natural, predominantly vertical shafts as a thin film of water flowing along their walls; (2) vadose flow or the flow through enlarged joints and fractures of the vadose zone with a predominantly vertical direction; (3) vadose seepage or very slow sinking through the smallest karst joints and fissures of the vadose zone, also with a dominant vertical direction. The quantitative contribution of each component changes from one section to another, but is dependent on the period of the year, vegetation cover, preceding precipitation, etc. If the karst is bare, the influence of the surface flow in the concentration phase is not significant; in addition, it can be frequently neglected. Lateral flow in the layer of non-consolidated soil occurs only in those regions where such soil is dominant. The overland flow can either occur or not depending on the infiltration characteristics of that layer in covered karst. The importance of the subcutaneous flow in the karst area is great; consequently, the following chapter deals with that problem. The vertical water flow through jamas significantly varies by its quantity in time (depending on precipitation), but essentially it does not quantitatively represent a significant factor, if a greater catchment area is considered instead of the narow area around the jama. It is difficult to distinguish quantitatively the flow through larger from the flow through the smaller karst joints and fissures. These two components have a dominant influence on the process of the vertical water flow through the vadose zone down to the groundwater level. Differences exist in the flow velocity as well as in the quantities varying in time. Whereas the water flows faster through larger fissures, and the hydrograph varies more evidently in time, the flow velocity in small joints is smaller (usually it is the laminar flow) and the hydrographs are practically constant.

A specific feature of the surface flow in karst is that open streamflows dry up in the dry period of the year. Drying most frequently occurs in summer, but it can occur also in the long dry winter periods. Drying is caused by the drainage of the stored groundwater quantities, and by the lowering of the groundwater level. One of the characteristic features of the groundwater flow in karst is the fast and considerable oscillation of the water level and great differences between the minimum and maximum groundwater levels. These oscillations can reach values of the order of 100 m and more. This fact shows that the karst water storage capacity is not so great as it appears at first sight. The water circulation in the vadose zone is, however, very fast and turbulent due to the size of fissures and a well-developed system of water circulation.

Generally, there are two types of water circulation in karst. The turbulent flow, similar to the flow in pipe systems, occurs in the upper layers of well-developed karst areas, the so-called conduit flow. In the lower layers of karst with dominant, fine, small joints and fissures the water follows the principles of Darcy's law, similar to a flow in a porous medium, and is generally called diffuse flow. Atkinson (1977) studied the water circulation in karst on the terrains of Mendip Hills Somerset in Great Britain. His investigations showed that the volume, and hence, the capacity of the pipe system is about 30 times smaller than the volume of the karst mass, but that 60–80% of the total quantity of water is quickly transported through the system of these channels. This phenomenon can be explained by the fact that the channel system has a much greater permeability capacity. The quantity of 54% of the annual discharge from the Cheddar Spring in UK is supplied by the quick flow. Quick flow comprises stream sink water and percolation from closed depressions (Atkinson 1977). The remaining part, i.e. 46% of the discharge is effected by base flow. It includes slow percolation from areas not drained by depressions and possibly groundwater leakage also. Although Atkinson's investigations are related to a relatively narrow region, certain values and conclusions can be of more general importance for the karst areas.

According to the order of magnitude from Cheddar Spring similar values have been determined in the case of the Klokun Spring situated in the Dinaric karst. A hydrograph analysis for an 18-year series (1965–1983) has been carried out for this spring located in Dalmatia, in the catchment of the Baćina Lakes (Yugoslavia). Figure 3.1 presents a hydrograph separation which has been attempted on a storm-by-storm basis. The hydrograph for Klokun Spring reveals a flushy behaviour very similar to surface streams. This division was not performed according to any strict rules; instead, it was carried out according to personal estimation which cannot affect the results more than ±5% to a maximum of ±10%. For the Klokun hydrographs it has been concluded that 75% of the annual discharge is quick flow, whereas only 25% is base flow. The mentioned values refer to an average of 19 years, and they vary significantly from one year to another, ranging for the quick flow from a minimum of 60% to a maximum of 89%. It should be stressed that there is an essential difference between the terms conduit flow and quick flow, as well as between the base flow and diffuse flow, although certain common features cannot be neglected. The base flow represents the water sinking underground by slow percolation, i.e. by slow seepage through smaller cracks, whereas the quick flow supplies the groundwater to the karst region by a quick percolation across well-developed depressions, pits, shafts and other karst forms or by fast sinking of water from the open streamflows, sometimes only temporary ones.

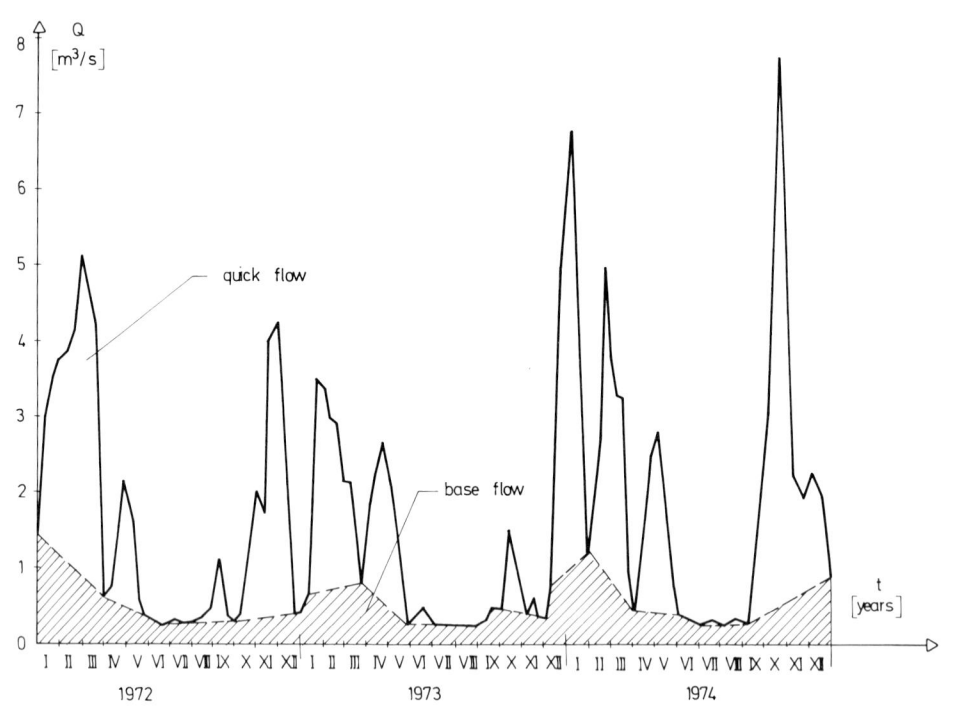

Fig. 3.1. Hydrograph of the Klokun Spring (Yugoslavia) in the time period from 1972 to 1974

From the point of view of engineering practice, it is important to note that in the karst aquifer the main part of the recharge to the exploitation wells is carried out by the base flow and the diffuse flow, since the quick flow, as well as the conduit flow, are too short to be significant for a greater and longer exploitation, particularly in the dry periods, when there is an increase in water demands.

The surface flow water in karst can gradually sink through a system of small joints and fissures, but sometimes all surface water sinks to the underground through a great swallow hole or through a system of several smaller swallow holes. Most often the above described situations occur simultaneously. E. White and W. White (1983) have concluded that the concentrated underground flows in karst maintain the profiles and gradients that they would have had if they had been flowing in surface channels. The conclusion reinforces the view that the groundwater flow in karst through privileged systems of channels (conduit flow) should be regarded as a "surface stream with roof" rather than as some sort of groundwater. The maintenance of these same or similar hydraulic profiles of the flow through the karst underground suggests that the development of the subsurface drainage, caused by a physical and chemical erosion of limestone, is a more rapid process than is the adjustment of the surface channels. The subsurface conduit system in karst seems to follow the development of the surface channels without difficulty.

There are three zones of water circulation in karst: (1) vadose, inactive or the zone of vertical circulation; (2) high water stand zone or the zone of horizontal circulation and (3) phreatic zone. The zone of vertical circulation is an area with large joints through which the water circulates quickly, predominantly in the vertical direction, down to the groundwater level. In that zone the water level increases quickly, exclusively immediately after heavy rains. The zone of horizontal circulation can be identified with a phreatic zone, i.e. the area where all the voids are filled with water. This book, however, does not deal with the chemical composition of karst water, or generally with the chemical agents used in karst investigations on water resources. It should, however, be emphasized that chemical analyses employed in karst investigations are extremely significant, particularly in studying the relationship between waters from the zone of vertical circulation entering the zone of horizontal circulation. Most frequently the chemical composition of water from the aerated zone of vertical circulation essentially differs from the water content in the zone of flooded karst. This is particularly evident after long dry periods. The zone of siphon circulation is most often found where there is contact between the karst and the sea. This zone has in general been least carefully studied and some authors believe its existence has neither been firmly proved nor does it present important evidence for the explanation of the processes of water circulation in karst.

The covering soil layer in karst, from the hydrologic point of view, has a limited influence on the formation of the processes of surface, subsurface and groundwater flows. Its existence of absence (without regard to its thickness and distribution) cannot significantly influence the change in the flow characteristics typical for karst. It has been discovered that the covering layer of non-consolidated soil above carbonate rocks makes chemical erosion stronger. This fact results in a greater number of better developed underground phenomena occurring in covered karst when compared to bare karst. The overland flow occurs rarely or hardly ever, in the areas of bare karst, whereas in covered karst it occurs regularly after heavy precipitation. This phenomenon can be explained by the essential difference in the infiltration capacity between bare and covered karst. The infiltration capacity of bare karst is great and practically does not decrease in time. In covered karst the infiltration depends upon the permeability features of the non-consolidated covered soil. However, when the water infiltrates through that less permeable layer it comes across a carbonate rock mass which is generally more suitable for water seepage than the rock mass in the bare karst. It is then that the water circulation typical for the karst terrains begins.

The aerated or vadose zone (Fig. 3.2A) is an area of continuous and quick groundwater level oscillations. The water circulation in the horizontal and vertical direction, as well as its appearance on the surface are primarily influenced by the development of the karstification processes in time, i.e. the process of the development of the surface and especially underground phenomena. This statement can be illustrated by an example showing the development of a spring location in time. As the base of the karst erosion was lowered in time, so was the location of the spring. The old spring passed from the zone of horizontal circulation into the zone of vertical circulation; it stopped functioning continually, and in time it completely dried up. It became exclusively a karst phenomenon with no special function for transporting water. Most often they become caves, and more rarely jamas (Fig. 3.2B).

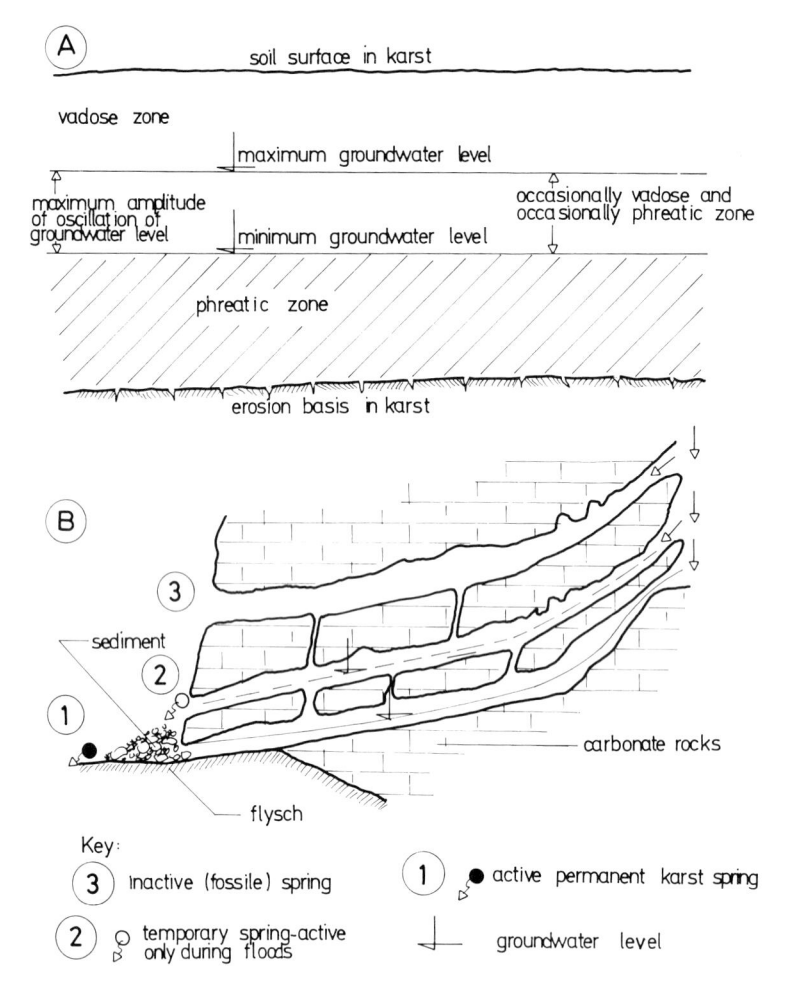

Fig. 3.2 A, B. Phenomena of groundwater in karst. **A** Vadose and phreatic zone; **B** historical development of spring levels in karst

The base of karst erosion changes constantly and the outflow level, as well as the groundwater level in a wider area, follow these changes, i.e. the spring location. When considering these phenomena we should distinguish the interest of geology, i.e. interest in the long-lasting evolution processes of karst phenomena in space and time, from the interests of practical engineering disciplines, such as hydrology, which is interested in the present conditions of karst to be applied in engineering practice. It is generally concerned with the construction of a structure to be used by people in a limited period of time, which is insignificant for geologic considerations. The purpose of these investigations is to make these structures function efficiently, ensuring a quick return of the invested financial means. In order to satisfy these requirements the characteristics of karst terrains should be studied in detail, particularly considering the water circulation in them. The best

Principles of Karst Groundwater Circulation

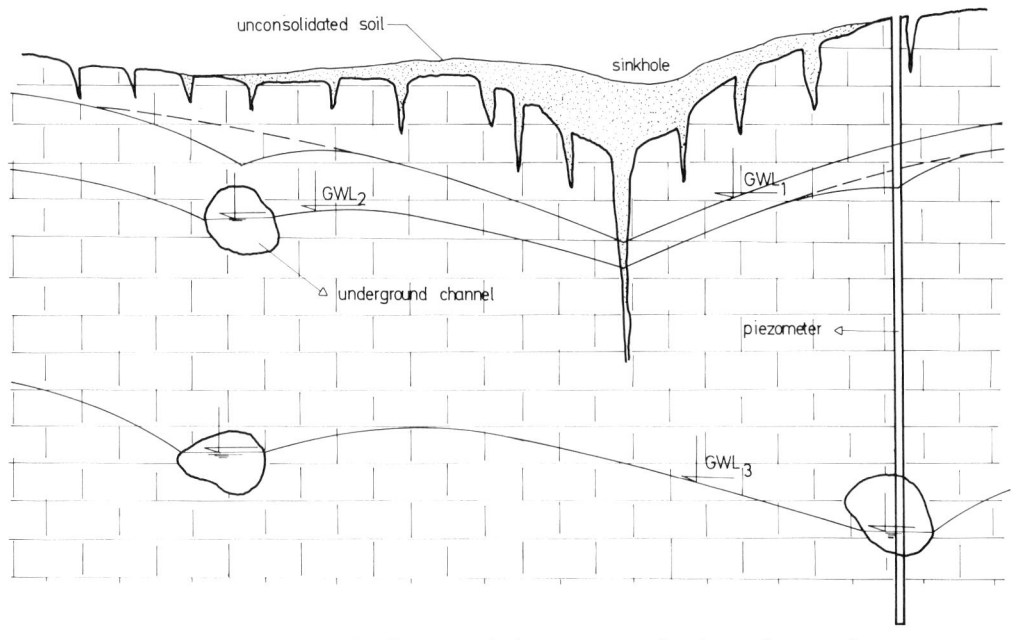

Fig. 3.3. Influence exerted by karst landforms and phenomena on the change in groundwater levels (GWL$_i$)

way to carry out such investigations is to establish a network of piezometers which can include natural piezometers as well (i.e. karst phenomena – jamas and caves).

Figure 3.3 presents the influence of underground karst phenomena on the change in the groundwater level. Various alternatives are so numerous that they are difficult to classify. Consequently, several important data on the groundwater level, obtained by piezometers, should be presented. According to Borelli (1966), the piezometric levels should include the predominant conditions of the groundwater flow in the greater part of the karst rock mass and provide more significant quantitative-qualitative data on the same process than the data on permeability. It should be stressed that the permeability measurements cost a great deal more than the measurements of the groundwater level by means of piezometers. It does not mean, however, that permeability should not be measured at all in karst terrains. Moreover, those measurements are indispensable, particularly in the design and construction of grouting curtains, but they should always be accompanied by groundwater level measurements, and by other types of measurements, primarily performed by geophysical methods. The local characteristics of the karst mass significantly vary in space; this frequently brings about the fact that some piezometers are influenced by local phenomena and do not display the features of the greater area. Figure 3.3 gives a schematic representation of this phenomenon, whereas Figure 3.4 presents the actual measurements of the groundwater level obtained by five piezometers in the Krčić River Catchment. Evidently all five piezometers show, in general, identical changes of the groundwater level, but those levels are significantly affected by the local characteristics

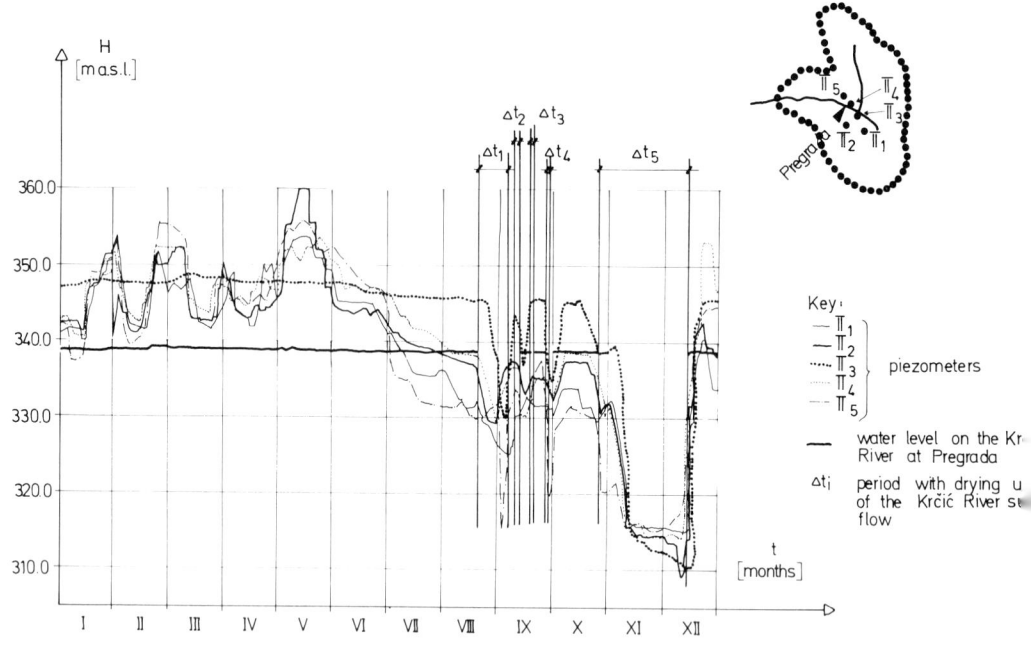

Fig. 3.4. Hydrographs of groundwater levels measured in piezometers in the Krčić Catchment in 1978

of the karst mass surrounding each piezometer. Borelli (1966) states that the groundwater level in a piezometer corresponds to the groundwater level surrounding it, only if the karst mass is permeable. If the recipient of the piezometer is not entirely separated from ist surroundings, then the groundwater level measurements represent an average piezometric level of all the layers through which it had been drilled. Furthermore, it is necessary to mention the frequent occurrences of a great number of suspended layers of groundwater in karst, which represent the function of the position and relation between the permeable and impermeable rocks in karst. Generally speaking, it can be stated that the permeability of the upper layers in karst is greater than the permeability of the lower layers. If this is true, it follows that it takes a shorter time for the piezometers to be filled than to be emptied, and this results in the fact that the groundwater levels shown by piezometers are higher than those occurring in the surrounding karst rock mass. This is particularly true for shorter periods of time with heavy precipitation occurring after long dry periods when the groundwater levels are very low. It happens quite often in the karst rock mass that groups of piezometers located in the vicinity do not show identical or stable groundwater levels. This phenomenon can be explained by the fact that the porosity of the karst rock mass is neither homogeneous nor isotropic, and can be additionally caused by the existence of isolating flows or voids with high permeability.

The most important data obtained by piezometers are those related to extreme water levels, both minimum and maximum ones. The prevailing opinion is that the minimum groundwater levels correspond to the height of the isolator location, and consequently are directly related to the change in the level of the erosion basis in karst. According to Borelli (1966), this explanation can be accepted only in special situations, and the minimum groundwater levels represent exclusively the levels reached by the groundwater after long dry periods. The same author, analyzing the data obtained by piezometers in Buško Blato, discovered the existence of a stable groundwater level near the maximum levels, characteristic for spillways. They can be explained by the existence of a strongly karstified zone through which the water flows with insignificant losses. In the case of Buško Blato these stable levels are caused by the functioning of the swallow holes situated on the boundaries of the polje. A similar situation was observed and described by Bonacci (1985) in the Krčić Catchment. Very important data, among those obtained by piezometers, are those related to the maximum velocities of the rising and lowering groundwater level. These data represent an important basis for the permeability analysis of the karst rock mass and its capacity for water storage, which will be discussed in the next chapter.

Table 3.1 presents essential data on the groundwater levels measured in boreholes in the Cetina Catchment, a typical Dinaric karst river, near the Prančevići Reservoir. These piezometers are quite deep, about 150 m. Evidently the maximum oscillation amplitude of the groundwater level ranges from 70 to 120 m. The maximum intensities of the groundwater level increase after heavy rains and vary from 1.3 up to 3.2 m h^{-1} or 36 to 77 m day^{-1}, whereas the maximum intensities of the level decrease in the period of droughts are significantly slower and range from 0.22 to 0.31 m h^{-1} (5.3 to 7.4 m day^{-1}). In the Krčić Catchment, located in the Dinaric karst, northwest of the Cetina Catchment, the maximum intensity of the groundwater level increase ranges from 0.8 to 1.7 m h^{-1} (20 to 40 m day^{-1}), whereas the maximum intensity of level decrease varies from 0.25 to 0.60 m h^{-1} (6 to 14 m day^{-1}); consequently, the values in both cases appear to be similar.

The complex and consequently complicated features of water circulation in karst make it difficult to either study theoretically the relation water-karst or to exploit wholly the otherwise abundant water resources in karst. Recently, both the investigation methods and the identification of the karst medium structures have developed rapidly together with the advances in the hydrologic, hydrogeologic and

Table 3.1. Groundwater levels measured in four boreholes in the Cetina Catchment near the Prančevići Reservoir (functioning from February 1977 – July 1979)

Number of boreholes	Maximum level [m a.s.l.]	Minimum level [m a.s.l.]	Maximum amplitude ΔH [m]	Maximum intensity of groundwater level increase [m h^{-1}]	Maximum intensity of groundwater decrease [m h^{-1}]
1	273.11	199.21	73.90	1.99	0.229
2	275.07	203.04	72.03	2.33	0.215
3	274.75	154.47	120.28	3.17	0.306
4	295.39	213.84	81.55	1.34	0.310

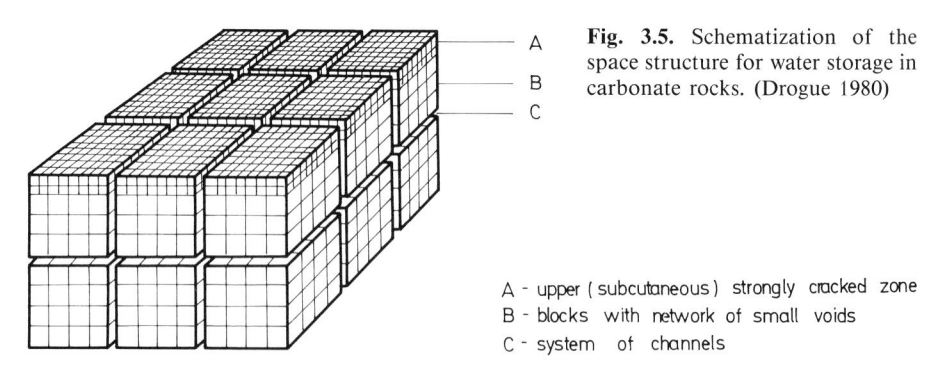

Fig. 3.5. Schematization of the space structure for water storage in carbonate rocks. (Drogue 1980)

A - upper (subcutaneous) strongly cracked zone
B - blocks with network of small voids
C - system of channels

hydrodynamic analyses: hence, the modelling of outflow processes in karst has progressed too. A simple model for the karst structure, applicable in the initial stage, was presented by Drogue (1980): its schematic representation is given in Figure 3.5. He states that when analyzing water flow in karst it is essential to describe, as precisely as possible, by a single schematization, the internal structure of the possibility for water storage in karst. It should be stressed that the spatial distribution of various permeability zones, and hence, their capacity in the karst system occurs almost accidentally and quite irregularly. This fact should not present a discouragement; on the contrary, it calls for a physically justified schematization of the outflow processes in the karst fissures and channels, and this should make possible a correct mathematical modelling of these processes. Until there is a detailed schematization of the karst medium structure, the outflow processes in karst should be studied only by probability and conceptual simulations. Such an approach should be overcome in the future. The scheme of the structure presented in Figure 3.5 ensures only the initial steps of that process.

In his schematization of the karst medium Drogue (1980) starts from the assumption that the existence of fissures in the carbonate rocks is an essential condition for the development of that medium with regard to the space for water storage, through which it can flow in the vertical or horizontal direction. The carbonate mass is most karstified in the areas of contact with the surface and just below it, since the water sinking underground in that place has the maximum potential of chemical and physical aggressiveness. The influence of the subcutaneous zone in karst and its karstification will be analyzed in detail in the next chapter, and its position in the structure model presented in Figure 3.5 has been designated by A. C expresses the flow in the irregular channel system of karst and B denotes the flow in the cracked blocks with a network of fine fissures. The distinction between the water circulation in those two zones is significant, and it can be schematically presented as in Figure 3.6. During heavy rains, the groundwater levels in the underground channels rise more rapidly than the levels in the surrounding karst mass, thus causing the water to flow from the channels into the karst mass. During long-lasting dry periods the water flows in the opposite direction. The groundwater levels in the channels are lower and they drain the water from the surrounding karst areas. It should be repeated, however, that the storage capacity of the channel system is small compared with the storage capacity in the network of small fissures. If the fissures are exceptionally small, a layer of

Principles of Karst Groundwater Circulation

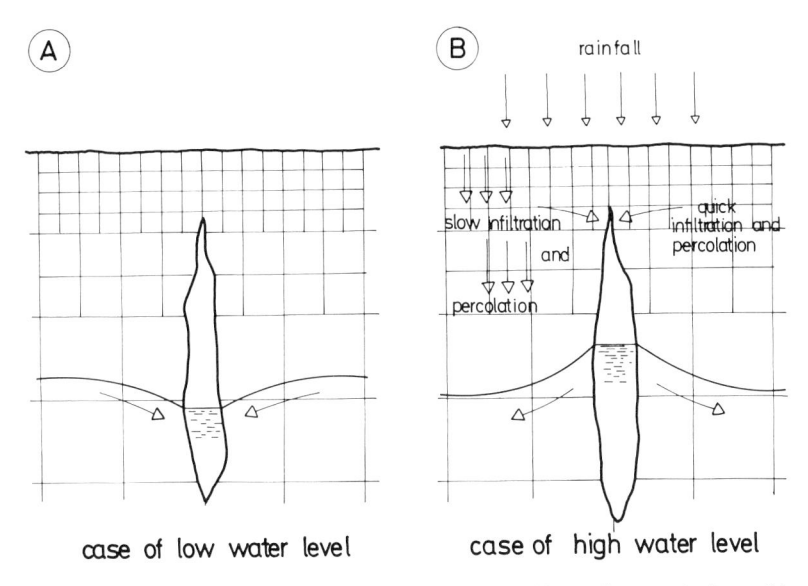

Fig. 3.6 A, B. Water circulation in the karst network of large fissures. **A** Case of low water level; **B** case of high water level. (Drogue 1980)

capillary flow from 0 to 50 cm deep can be formed above the groundwater level. The influence of zones B and C on the outflow in the structural model is represented in Figures 3.7 and 3.8. The differences between the changes in the groundwater levels obtained by two neighbouring piezometers, due to the existence of zones B and C, are also presented. Piezometer T_1 is located in zone C and it responds rapidly to pumping, whereas piezometer T_2, although close to the other, responds slowly. Figure 3.8 presents the different reactions of the groundwater levels to rainfall of equal intensity in two closely situated piezometers. These different reactions can be explained by their position in the karst mass. The water from both piezometers appears in spring A. While piezometer T_2 is located in the system of developed fissures and channels, ensuring rapid transfer of water to the springs, piezometer T_1 is situated in a system of small fissures and thus its reaction is slower. This example illustrates the fact that one should be careful when dealing with the velocities of the water circulation in karst. Consequently, the flow and velocity of water circulation in a developed fissure system should be distinguished from that in a system of channels. Drogue (1980) emphasizes that the ratios water velocity-gradient, obtained by numerous measurements, can be 1000 times smaller in a network of fissures than in a system of channels. In order to support the suggested structure model, Drogue (1980) shows the influence of the karst channels with different thermic profiles on two close piezometers, as presented in Figures 3.9. The differences in permeability between zones B and C have direct thermic consequences. In this case the existence of the underground channel and the direct, rapid circulation of cold water have cooled the groundwater in the piezometer. Understandably, the water can be heated locally if the water flowing through the channel is warmer than the water in the surrounding

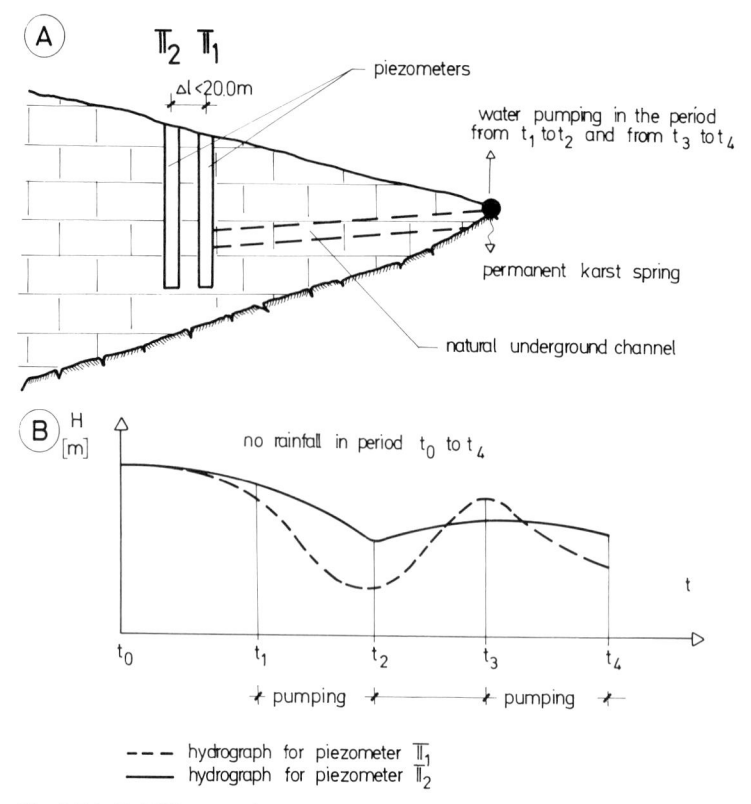

Fig. 3.7 A, B. Differences in changes of groundwater levels **B** in two piezometers in media of different permeability **A** caused by water pumping of the spring. (Drogue 1980)

karst mass. The great velocity of the water flow in the system of channels and its short retention in the underground make it impossible for the massive to affect significantly this water by its thermal potential. The model presented by Drogue (1980) did not completely succeed in describing all the complex features of the karst medium functioning and affecting the flow process. Thus, the random element present in practice is missing. The model has no general significance, but it represents a sound physical basis which should be developed and modified for the actual situation in various types of carbonate reservoirs.

3.1 Subcutaneous Zone (Epikarstic Zone)

The Drogue structure model for karst and some earlier sections of this book stress the enormous importance of the surface and immediate subsurface zone (Fig. 3.5A) for the karstification processes and for the water circulation in karst. This first immediate contact between the carbonate mass and atmospheric water is important enough to be carefully studied. The technical term in many languages is subsurface zone, while the English use a more precise term, i.e. subcutaneous zone.

Subcutaneous Zone (Epikarstic Zone)

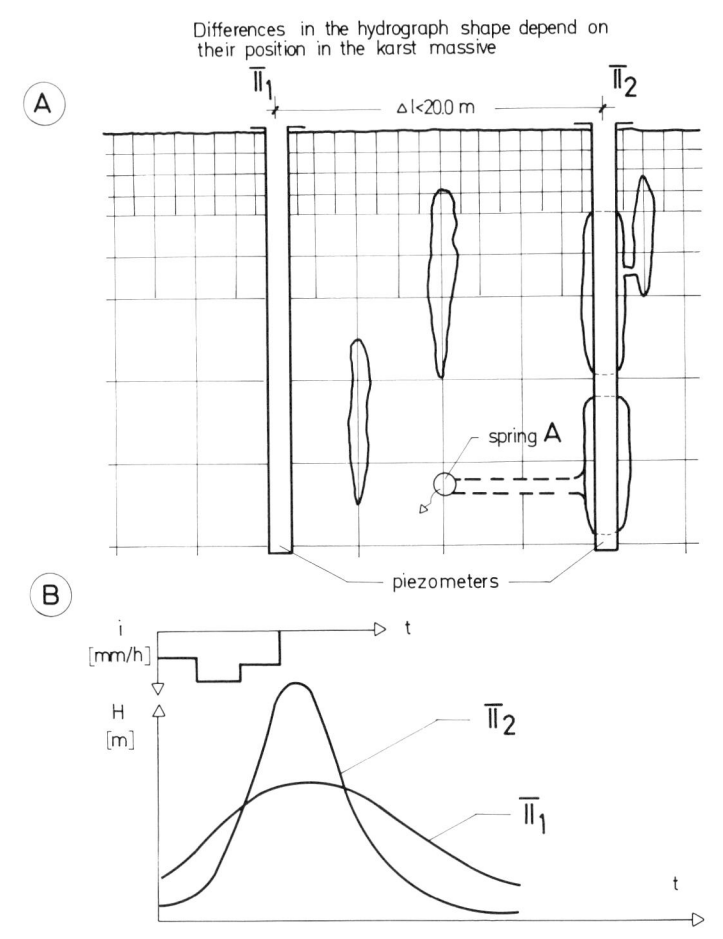

Fig. 3.8 A, B. Hydrographs of groundwater **B** in two neighbouring piezometers **A** caused by the same rain event. (Drogue 1980)

Williams (1983) describes the role of the aerated and non-saturated subcutaneous zone where the water flows vertically and reaches the groundwater level as having the following characteristics: (1) it most often represents a large reservoir where the water can be stored for a certain period of time; (2) the water flow within the zone, very often displays, particularly after heavy rainfall, a significant lateral component; (3) this relatively narrow zone plays an important role in the groundwater inflow, although its role is often underestimated compared with the role played by the zone located below it and stretching to the groundwater level. It is almost unnecessary to emphasize that the storage features and the lateral component of the water flow in the subcutaneous zone are not even approximately uniformly distributed within the zone. A more precise definition of the subcutaneous zone includes a relatively narrow region of an average depth ranging from 0.5 to 2 m, located under the non-consolidated soil (if it exists) at the very top of the non-saturated zone, in the area with the best-developed, current,

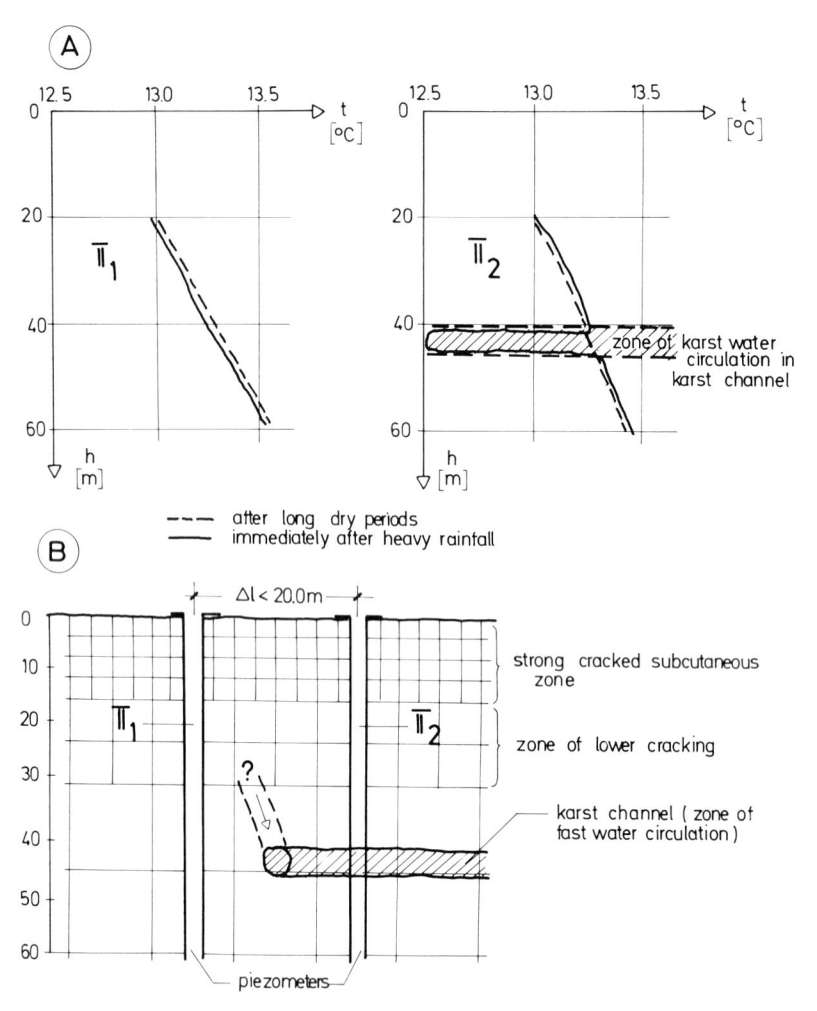

Fig. 3.9A, B. Thermic profile **A** in two piezometers **B** dependent on their position in the karst. (Drogue 1980)

karstification processes. It is, thus, a zone of carbonate rocks, directly affected by climatologic factors. This zone is primarily important for the development and explanation of hydrologic processes, as presented by Drogue's structure model (1980). In geologic and hydrogeologic publications this zone is referred to as an epikarstic zone, used to stress the presence of numerous fissures, channels and tubes used for the rapid water flow or for the temporary storage of water.

The subcutaneous zone exists practically everywhere, both in bare and covered karst regions. One example is presented in the photograph of a Srinjine Quarry near Split (Dalmatia, Yugoslavia) (Fig. 3.10), and similar situations can be found in any natural or artificial cutting. The photograph shows that the rock cracks are best developed in the top zone ca. 1 m deep. At a distance of every 10 to 50 m there are vertical fissures, reaching a depth of 30 to 90 m. Although this case refers

Subcutaneous Zone (Epikarstic Zone)   31

Fig. 3.10. Photograph of a vertical section through the karst massive in the Srinjine, Quarry (taken by Granić)

to bare karst, all or many of the fissures of the subcutaneous zone, as well as the other vertical fissures, are filled with fertile non-consolidated soil. This area is covered by terra rossa, being the only fertile soil in a wider area, with Mediterranean vegetation, very resistant to severe climate, i.e. wind and drought. Williams (1983) explains the origin of this zone primarily by solution. The water sinking underground contains CO_2 and this results in the formation of carbonate acid which decomposes carbonate rocks. As the water sinks deeper, the corrosion decreases due to the reduced quantity of CO_2 in water. The quantity and depth of chemical and other types of erosion, depend on the quantity of rainfall, its distribution in time, sinking (infiltration and percolation velocity), its composition, the thickness of the soil and rocks, partial pressure of CO_2, vegetation and other

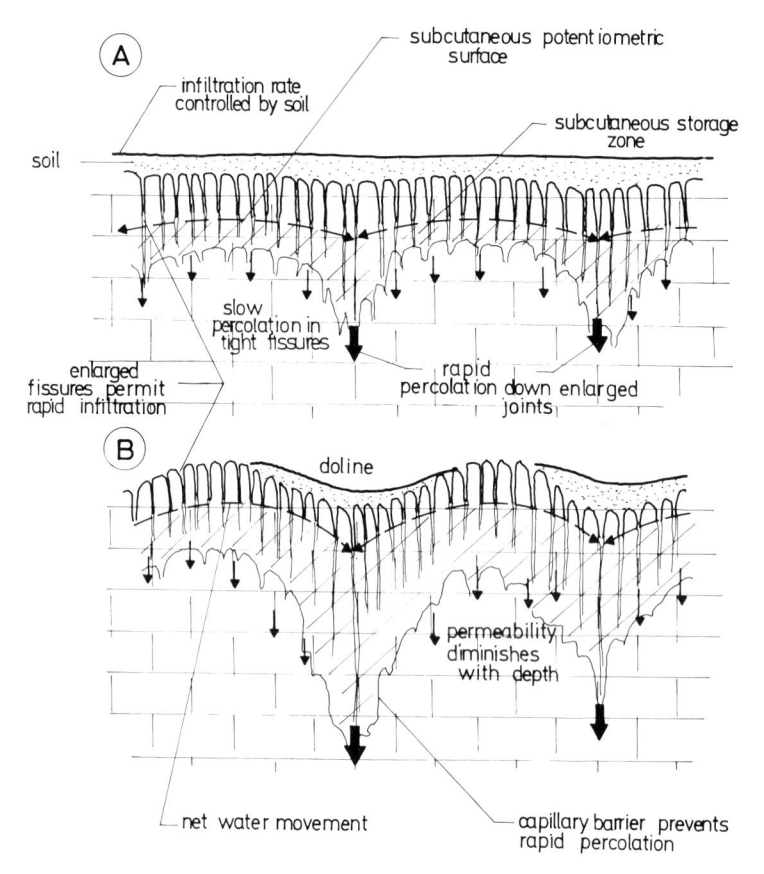

Fig. 3.11 A, B. Subcutaneous storage, lateral subsurface flow to zones of high permeability **A**, and the evolution of solution dolines **B**. (Williams 1983)

parameters. Kogovšek and Habič (1980) state that in a vertical trickle in the Planinska Cave they discovered that 42 m^3 water dissolved ca. 7 kg carbonate rocks during 17 h and transported to the jama 6 kg of suspended matter. These processes are exceptionally strong and they currently change the surface of the karst terrains. Figure 3.11 depicts the influence exerted by the subcutaneous zone on the vertical and lateral water circulation and the formation of sinkholes (Williams 1983). Thus, the important role of the non-consolidated soil gathered in fissures should be stressed, not only considering its influence on chemical erosion, but also regarding the regulation of infiltration and percolation, and its functioning as the storage of water on its way to the phreatic zone. The water flow through the subcutaneous zone regulates the water inflow to the caves. This flow essentially differs from the base groundwater sinking, as the water does not flow from the saturated phreatic zone, but from the non-saturated vadose region.

The previously stated facts make it necessary to analyze the vertical trickle transporting the water in the fastest way and regulating the flow from the subcutaneous zone to the karst mass, particularly in the system of caves or

Subcutaneous Zone (Epikarstic Zone)

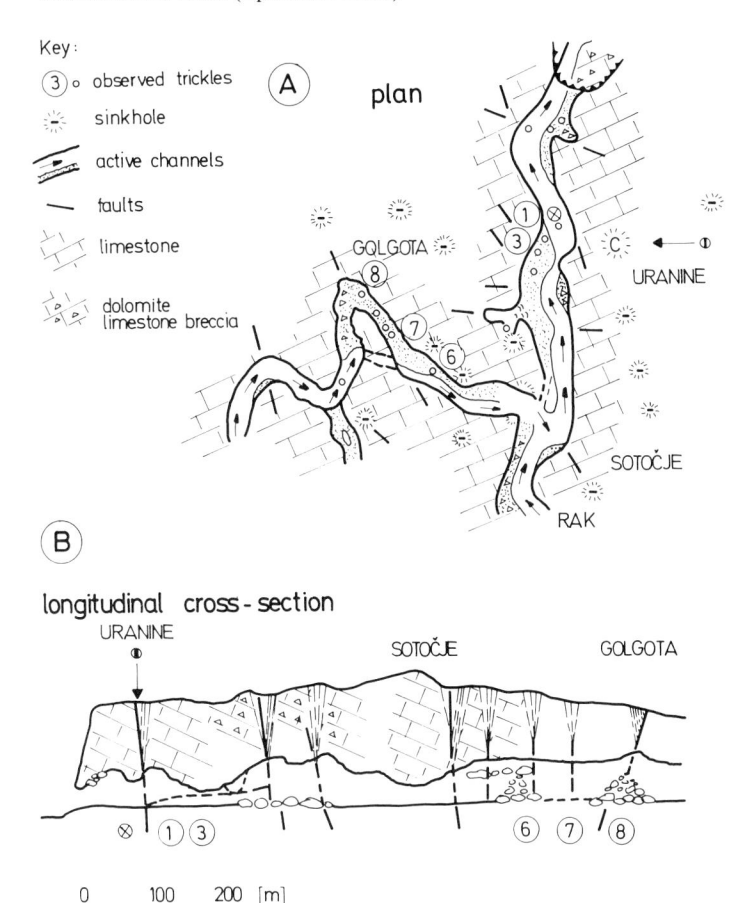

Fig. 3.12 A, B. Vertical water percolation in the Planinska Cave (Yugoslavia). **A** Plane; **B** Longitudinal cross-section through the cave. (Kogovšek and Habič 1980)

underground karst openings. The problem has been intensively studied on the system of the Postojnska and Planinska Caves by researchers from the Institute for karst investigation SAZU from Postojna, Yugoslavia. Figure 3.12 presents a situation and cross-sectional area of the Planinska Cave with plotted vertical fissures (trickles) used for measuring the water sinking into the jama. This was taken from a paper by Kogovšek and Habič (1980). These authors tried to determine the manner and velocity of the water sinking, and the precise trajectories of the water flow to the top of the cave vault by tracing methods, using uranin, and by adding water to the sinkhole surface during drought periods. There are permanent and temporary trickles, bringing water only during the rainfall and some time after its cessation. The water quantity delivered by a single trickle was measured in the cave using a measuring glass and a stopwatch; apart from the water quantity, both its temperature and the temperature of the air were measured. The catchment area was determined for each analyzed trickle primarily in the dry period; these areas are quite small, from several square meters to a maximum of 1000 m². The drainage system of the trickles is composed of several differently pervious conductors, where the water could be retained. The drainage systems of neighbouring trickles are mixed with the others. Minimal and maximal discharges of different trickles occur at the rate of 1:100 up to 1:1000 or even 1:5000. The extent of the drainage system influences the annual trickle capacity; therefore, the annual water quantities of different trickles can occur at the rate of 1:100. This seems to be an essential property of karst surface dissection. The percolation occurring through a ca. 100-m-thick cave ceiling varies considerably in space and in time. In places with continuous water dripping the conditions favour the formation of stalactites. Trickles with greater water quantities result in larger leaks, or have an opposite effect, i.e. the calotte surface is throughly washed off. The flow in smaller trickles is more constant, whereas the flows through the greater ones significantly vary in time. The maximum discharges obtained by measurements reach $15000 \, l \, h^{-1}$.

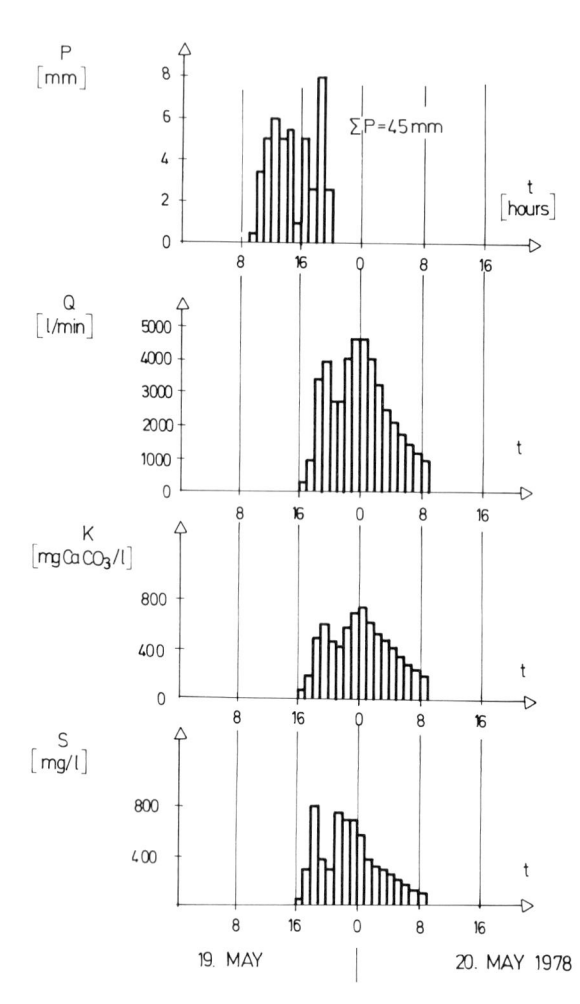

Fig. 3.13. Measurement data on precipitation (Q) carbonate hardness (K) and concentration of suspended sediment (S) on trickle 1 in the Planinska Cave (Yugoslavia). (Kogovšek and Habič 1980)

Trickles do not respond to all types of precipitation. After long-lasting dry periods, depending on the air temperature and vegetation, the water will manage to sink from the surface to the calotte of the cave only if the rainfall is of the order of 70 mm, whereas in the wet period even the water resulting from rainfall of 5 mm manages to penetrate. These data refer to the significant storage capacity primarily of a subcutaneous zone, and subsequently of the remaining non-saturated part of the massive. The experiments by tracing have efficiently proved that the water flow trajectories are neither vertical nor direct. The trickle hydrograph is smaller than the rainfall pluviograph as presented in Figure 3.13. The same process refers to the concentration of suspended sediment and to carbonate hardness. Corrosion and erosion directly depend on the quantity of percolated water, and the karst denudation is by far the greatest below the system of sinkholes situated above jamas. Kogovšek (1982) presents the results obtained by measuring the discharges in trickles 1 and 6 and the respective rainfall quantities. The results are presented in Table 3.2. Evidently, more abundant yearly precipitations result in greater water quantities. The data related to 1980/81 appear illogical, since small water quantities result from relatively heavy rainfall. The reason why this is so is that the measurement periods cover the time from March to February of the following year. In January and February there was much snow which did not thaw and so this water could not enter the process of infiltration and percolation to the cave vault. It was noted that in 1980/81 about 50% of the total water quantity managed to sink during a very short period, i.e. 55 days. The regularity of the water distribution in time, flowing to the cave vault depends both on the precipitation distribution in time and on the evapotranspiration intensity.

Williams (1983) studied the delay in the rise of the level in the small lakes of the Carlsbad Cave resulting from the water percolation through the karst mass ca. 250 m thick. The hydrograph responds to precipitation with a delay of 2 to 14 weeks, thus witnessing the enormous importance of the water storage particularly in the subcutaneous zone. The same author stresses the existence and significance of a capillary barrier at the bottom of the subcutaneous zone. This barrier regulates the percolation into the deeper vadose layers, since the water pressure from subcutaneous storage must exceed surface tension or capillary forces in narrow fissures before percolation can occur, and only in wider fissures are surface tension effects relatively insignificant (Williams 1983).

Table 3.2. Annual quantities of rainfall and water flowing through trickles 1 and 6 in the Planinska Cave. (Kogovšek 1982)

Ordinal No.	Year	Rainfall	Water quantity [m^3] Trickle 1	Water quantity [m^3] Trickle 6	Discharge [l h^{-1}] Trickle 1 min	Trickle 1 max	Trickle 6 min	Trickle 6 max
1	1977/78	1882	1900	–	1.2	12000	10	–
2	1978/79	2097	2000	–	1.2	4200	5	6300
3	1980/81	1976	1100	1100	3	4800	3	6600

3.2 Karstification Depth and Karst Capacity for Water Storage

The objective of studying the karstification depth and the karst capacity in this book is related to the needs of engineering practice, i.e. in making engineering decisions concerning the possibility and efficiency of using the water resources in an area. Sometimes, the problems deal with the spring intakes, necessity for grouting or other procedures to reduce or even completely eliminate the losses, and sometimes the problem is defined as the formation of a groundwater reservoir in karst. From the hydrologic standpoint it is always the same problem eventually related to the necessity of defining the water quantity in space and time. This space is both non-homogeneous and located under the surface. The hydraulic-hydrologic approach has definitely proved to be impractical since it refers to the homogeneous porous media where the water flow is predominantly laminar. To determine the water quantity in the karst underground it is necessary to define the karstification depth and the change in its quantitative values at different levels of depth. Unfortunately, this problem cannot be solved by a single scientific discipline; furthermore, it can be stated that an interdisciplinary approach does not succeed in giving a precise solution to the problem even in the simplest cases. The importance of hydrology in solving these problems can be primarily stressed in using data on surface flows and springs and when studying the variations of water quantities in time and space, so that they can be properly exploited.

The problem of karst capacity for water storage has already been dealt with in the literature. Numeric data exist, which all have common characteristics, i.e. they were obtained by different methods and all the authors emphasize that they represent only approximate estimations. We have to accept the fact that the results are approximate and will remain so for a long time. However, by analyzing the methods we can discover possible errors in the estimation of karst capacity and void volume in karst. Primarily, the void volume to be considered in our analysis should be defined. For practical purposes it is most convenient to define the effective porosity n_e as the pore values (expressed in percentages) located in the area between the maximum amplitude of the groundwater level oscillations. This definition cannot be accepted, however, without additional comments as illustrated in Figure 3.14. The effective porosity should refer to the part of those pores in the karst rock mass from which the air can be pressed out by water and which are found within the area of the maximum amplitude of the groundwater oscillations. It should be noted, however, that a certain quantity of air always remains within the massive. This explanation is not quite correct as illustrated by Figure 3.14A. The effective porosity, for engineering practice, should refer to the area of dynamic, stored-water quantities in a natural reservoir, whereas our interest in the effective porosity in the zone of static storage should be less important until these stored quantities become exploited by a structure (e.g. a tunnel or a gourd for siphoning). The situation becomes more complex when the effective porosity has to be defined for the ascending spring as schematically presented in Figure 3.14B. In front of the spring there is a significant impermeable barrier of great thickness. Such a spring is Bulaž in Istra (Yugoslavia). Figure 3.14C illustrates the case where there are no static, stored-water quantities. This situation

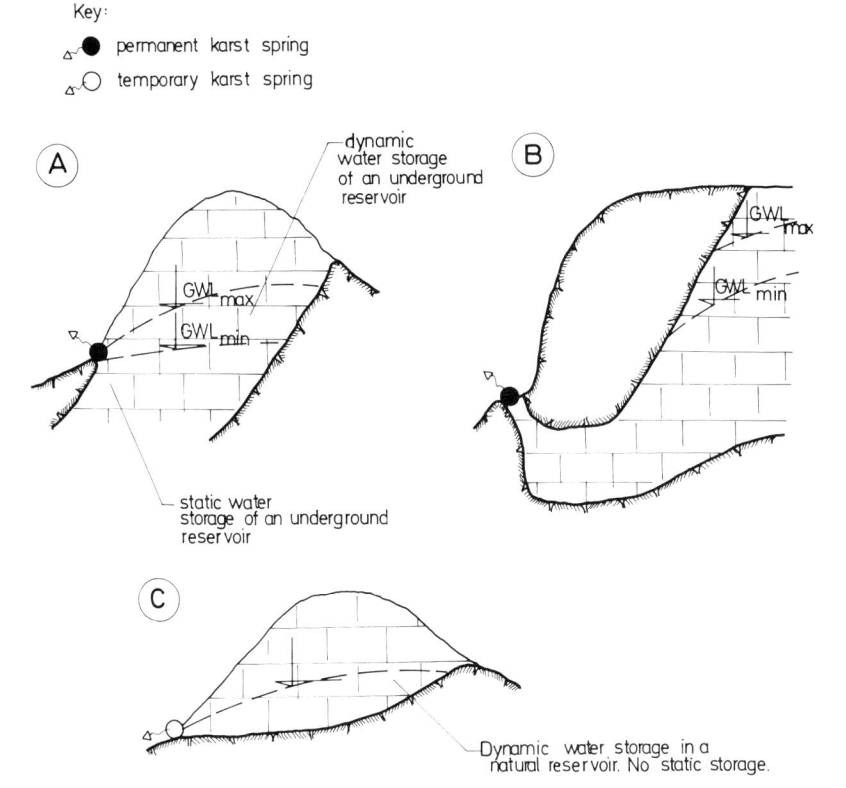

Fig. 3.14 A–C. Explanations connected with the determination of effective porosity. Dynamic and static water storage reservoir, for descending (**A**) and ascending (**B**) springs. **C** Spring without static water storage

occurs in shallow karst. In order to define the actual maximum amplitude of the groundwater level oscillations, a long series of observations should be carried out (covering at least 50 years). It should be noted, however, that all the previously mentioned facts do not thoroughly explain the problem of effective porosity, since the pore volume filled by water is the function of the dynamic conditions of the water circulation in karst and changes in time, since the karstification processes are still going on simultaneously with the process of colmation. All the previous statements point to the conclusion that all analyses can give only approximate solutions which are, however, precise enough for the needs of engineering practice.

The karstification depth varies from one region to another. The existence of karstified rocks was established by borings even at a depth of 2300 m (Milanović 1981). There is no strict difference between karstified and non-karstified rocks; it is a transitory zone, varying in depth, with no significant sinking below it. This is the definition for the height of the location of the erosion basis, for the karstification base or generally the beginning of the impermeable layer. It has already been stated in the literature that there is an absolute relation of proportionality between karstification and depth. It has been confirmed by the following authors: Milanović (1981); LeGrand and Springfield (1973); Atkinson (1977); Vlahović (1983); Drogue (1980); Borelli (1966); Burger and Pasquier (1984) and many others.

Motyka and Wilk (1984) measured the dimensions of fissures and karst channels in the karstified Triassic rocks in the surroundings of Olkusz in Poland. The measurements included width, length and the spatial orientation of the fissures in the underground mine workings situated 150 m under the surface, as well as the mentioned geometric elements on the surface. The most frequent width equalled 0.2 mm. The maximum values of width decrease with depth. On the surface they reach up to 35 mm, and at the depth of 100 m the maximum width is considerably smaller, i.e. it was up to 5 mm, and at a depth of 150 m it was up to 2.5 mm. Motyka and Wilk (1984) defined on the stated location the surface fissure index A_s, the surface density of fissures D_s and the hydraulic equivalent fissure width b_h. The surface fissure index A_s is expressed in percentages and is defined by the expression:

$$A_s = \left(\sum_{i=1}^{n} l_i b_i\right)/A , \qquad (3.1)$$

where A represents the cross-sectional area of the analyzed karstified rock segment in m^2, l_i is the length of a respective fissure in m, b_i is the width of the fissure in m, whereas n expresses the total number of fissures in the considered cross-section A. The surface density of fissures D_s is expressed in m^{-1} and is defined by the expression:

$$D_s = \left(\sum_{i=1}^{n} l_i\right)/A . \qquad (3.2)$$

The measurement results of the fissure width and length were used for the calculation of the hydraulically equivalent fissure width b_h expressed in m, and calculated by the following expression:

$$b_h = \sqrt[3]{\left(\sum_{i=1}^{n} b_i^3 l_i\right) / \left(\sum_{i=1}^{n} l_i\right)} . \qquad (3.3)$$

Table 3.3 presents the range of values for A_s, D_s, and b_h measured in the Triassic carbonate rocks in the Olkusz region (Poland) presented by Motyka and Wilk (1984).

The dimensions of fissures in karst vary significantly, ranging from the smalles or micro-dimensions to the largest with dimensions reaching up to 100 m. All these fissures serve for the storage and transport of water. In the literature there are numerous classifications with regard to their origin and size. From the standpoint of engineering practice, the main disadvantage of all these classifications is that they are too complex and difficult to apply in practice. Therefore, a simple and more general classification into three categories has been suggested: (1) pore P (pore space); (2) fissure, F; (3) channel, C. The quantitative definition of their dimensions is a very complex problem. Consequently, they can be divided according to the dimension of the change in the cross-section d. In pores $d<0.1$ mm, for fissures $(0.1 \text{ mm}) \leqslant d \leqslant (1.0 \text{ cm})$ and for channels, the radius is $d>1.0$ cm.

Table 3.3. Extreme values of A_s, D_s and b_h observed within the Triassic carbonate rocks of the Olkusz region (Poland)

Depth below the surface [m]	A_s [%]	D_s [m^{-1}]	b_h [m]
Terrain surface = 0	0.13 – 4.4	3.95 – 15.98	0.23 – 14.0
100	0.0046 – 0.30	0.21 – 3.58	0.60 – 4.48
150	0.070 – 0.66	2.46 – 14.68	0.56 – 1.03

Milanović (1981) suggests an experimental expression as the basic law of karstification development directly proportional to the depth as measured from the surface. It is the following expression:

$$\varepsilon = a \cdot e^{-bH}, \tag{3.4}$$

where ε is the karstification index expressed as the dimensionless quantity, H is depth in m measured from the surface, whereas a and b are parameters obtained by processing measurement data on soil permeability using the least squares method. It is an empirical expression valid only on the vertical or in a narrow region for which it has been determined. Milanović (1981) obtained, for the area of East Herzegovina (Yugoslavia), some parameters stating that karstification in the 10-m-deep layers is 30 times greater than at a depth of 300 m. Stepinac (1979), using a different approach, i.e. analyzing recession curves, concluded that the effective porosity at a depth of 300 m is 0.57%, whereas at a depth of 10 m, it is 14.18%, i.e. about 25 times greater. He states that these data, as well as the whole curve, refer to the average condition of the pore volume in the Dinaric karst, assuming the same height of an ideal prism.

Burger and Pasquier (1984) obtained the results on the change in the filtration coefficient k expressed in m s^{-1} and measured at different points in the function of various depths. These measurements were carried out in the Swiss Jura near Neuchâtel. The results are presented in Figure 3.15. Evidently k changes depending on the location, and the filtration coefficient decreases to a certain depth (150 – 300 m), and subsequently remains the same. Karstification can be qualitatively determined in many ways, but the actual quantitative values of the effective porosity are obtained only after laborious processes. The permeability studies, no matter how complex and expensive, do not yield either reliable data or a quantitative presentation of the porosity condition of a region.

The karst medium is very heterogeneous and most surprising. It displayed such characteristics also during the excavation of the Zakučac 2 Tunnel in the Cetina River Catchment (Yugoslavia). Bojanić et al. (1980) report that, along the stretch of 8.5 km, six large speleologic phenomena have been investigated, whereas some smaller ones have not been studied at all; instead, they were filled by concrete. Figure 3.16 presents two greater caverns on one of the tunnel profiles, stretching from the new to the old tunnel, used to discover damage on the old tunnel lining. The previously described situation is not rare in the Dinaric karst. In addition, it almost as a rule occurs during most underground construction operations. According to Milanović et al. (1985), while boring a tunnel 8.1 km long with

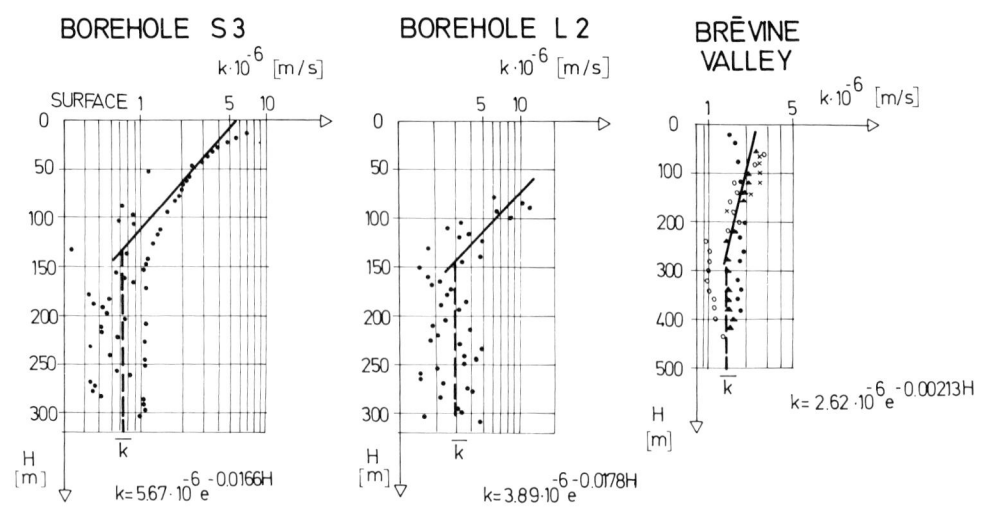

Fig. 3.15. Change in the seepage coefficient (K) measured in limestone areas in the Swiss Jura near Neuchâtel in the function of a depth (H). (Burger and Pasquier 1984)

a diameter of 8 m for the construction of PHP Čapljina (Yugoslavia), a karst cavern was discovered with partly clayey-sandy deposits and blocks of crushed limestone. Therefore, the tunnel required additional supports, and the critical zone was spanned by a specific vault structure. During the tunnel's exploitation water with a high concentration of suspended clayey particles appeared in the engine room on the pilot boreholes and in the nearby springs. This phenomenon occurred after a heavy rainfall of 117 mm day^{-1}. Great quantities of material were transported from the cavern, so that the tunnel remained suspended in the air along the stretch 15 to 20 m. The lateral and longitudinal cross-sections of the damaged part of the tunnel are presented in Figure 3.17.

The presented examples indicate that the permeability measurements are usually restricted to the narrow zone around the borehole, and that the conditions change in time. Furthermore, the flow regime is not known. It can change from one layer to another, and even within a single layer 5 m deep. Borelli (1966) studied the ratio between the filtration coefficient k expressed in m day^{-1} and permeability expressed in Lugeons (Lu), and he concluded that:

$$1 \text{ Lu} = 0.007 - 0.015 \text{ m day}^{-1} \, . \tag{3.5}$$

The same author studied the interaction between the specific permeability q expressed as l/min·m·0.1 atm and the filtration coefficient k suggesting for practical purposes the following relation:

$$k = (1.2 - 2.3) \times 10^{-5} q \, , \tag{3.6}$$

where k is given in m^{-1}s, and for approximate computations he suggests the application of the value of the constant 1.7×10^{-5}.

Babushkin et al. (1984) carried out investigations on water quantities in deep mineral mines of the North Ural (USSR). The mines are located in the karstified

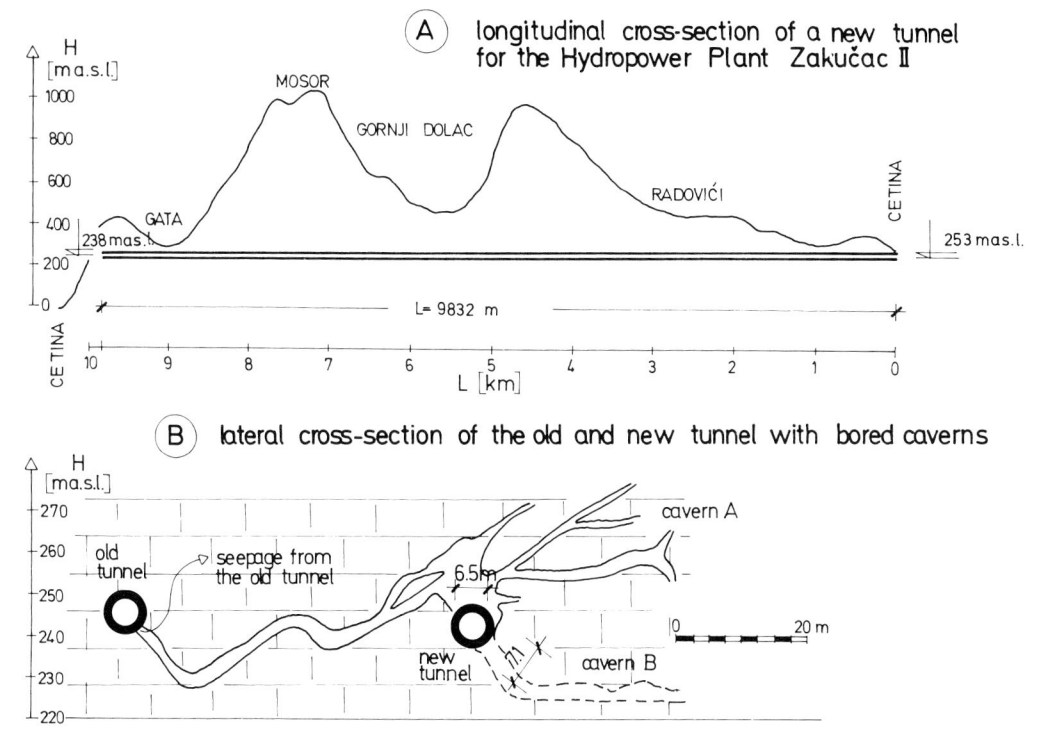

Karstification Depth and Karst Capacity for Water Storage

Fig. 3.16 A, B. Longitudinal (**A**) and lateral (**B**) cross-section of the largest caverns on a profile of the tunnels I and II. (Bojanić et al. 1980)

carbonate rocks, reaching a depth of 1000 to 1300 m. According to numerous investigations, they defined the analytical expression, representing the ratio between the rock karstification K_s and the permeability coefficient K, by the following expression:

$$\log K_s = 0.65 \log K + 0.34, \tag{3.7}$$

where K_s is the number of karst pores in 1 m of the bored core. The regression equation is valid for the local conditions of the considered region, and the correlation coefficient is 0.75.

All the previously mentioned facts point to the conclusion that permeability measurements alone are not sufficient to obtain a reliable definition of the effective porosity of a karst massive. It can be approximately defined in four various ways, which will be discussed later on. The first method refers to the groundwater levels measured on numerous piezometers in a wider region supposing the hydrologic boundaries of the catchment can be defined at least approximately. It includes the measurements of the maximum velocity of the rising groundwater level. In this stage, it is supposed that only the vertical water circulation affects the increase of the level, since the horizontal flow component cannot considerably contribute to the groundwater recharge. The expression for defining the effective porosity n_e is the following (Borelli 1966):

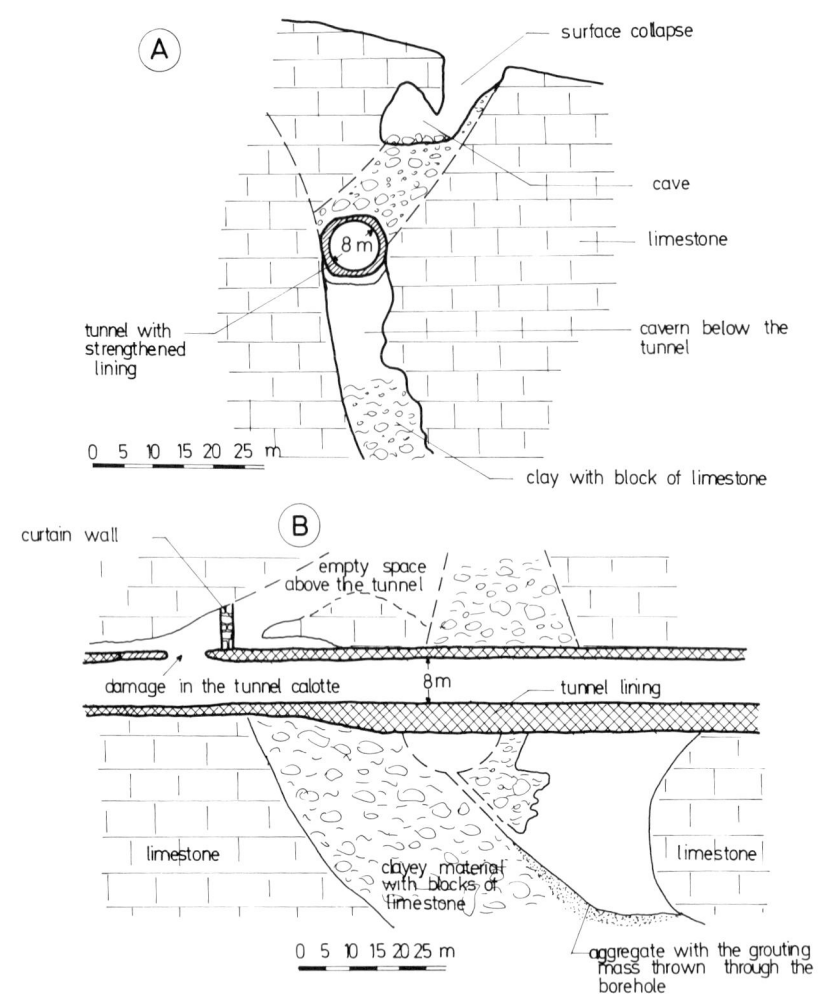

Fig. 3.17A, B. Lateral (**A**) and longitudinal (**B**) cross-section of the tunnel of the power plant Čapljina at the cavern in a karst tunnel. (Milanović et al. 1985)

$$n_e = (i \cdot I)/v_{max} , \tag{3.8}$$

where the intensity of the rainfall is i in m day^{-1}, I is the sinking coefficient, whereas v_{max} is the maximum velocity of the groundwater level increase in m day^{-1}. Although expression (3.8) is very simple, it should be very carefully used. The distribution of intensive rainfall in space and time is irregular, and the influence of the preceding state of moisture on the sinking coefficient is significant. In bare karst the recommended values for I range from 0.6 to 0.9. It is not recommended to analyze the data obtained by one piezometer or a small number of piezometers; instead, a wide analysis of the whole region should be carried out referring either to precipitation, or to the state of the previous moisture, and particularly to the velocity of the water level increase. If such an analysis is not carried out, the results will refer only to the effective porosity in the neighbourhood of the measurement point and cannot be considered representative for a wide region.

The values of effective porosity were computed for four deep piezometric boreholes in the Cetina Catchment (Yugoslavia), whose essential data are presented in Table 3.1. These values were computed when the horizontal flow component was considered negligible with regard to the vertical component of the groundwater flow. Depending on the chosen value of the sinking coefficient I, being in this case closer to the upper limit, since it is valid for an area of bare karst with well-developed surface and subsurface forms, the effective porosity values ranging from 1.5 to 3.6% were obtained. In all the cases I was taken to be 0.8. The differences between boreholes were not significant, and the lowest values were obtained in borehole 3, with the strongest intensity of level increase and decrease. According to data obtained by the above four piezometers, it can be concluded that the average effective porosity in that region of the Cetina Catchment (Yugoslavia), obtained from Eq. (3.8), is 2.5%.

The second method for determining the effective porosity is based on using the decreasing part of the hydrograph in the dry periods of the catchment, or, generally speaking, by using the recession curves. This method is primarily recommended for analyzing the effective porosity of the spring hydrographs in karst. This procedure for computing the effective porosity and the volume of the underground storages was used by: Stepinac (1979); Ivanković and Komatina (1976); Forkasiewicz and Paloc (1967) and many others. The following experimental expression was defined by Maillet (1905) for practical purposes:

$$Q_t = Q_0 e^{-a(t-t_0)} = Q_0 e^{-(\lambda/A)(t-t_0)} \,, \tag{3.9}$$

where Q_0 is the discharge in time t_0, Q_t *is the discharge in time t*, whereas a is the outflow coefficient which varies with the change in the microregime of the outflow (Fig. 3.18 A, B, D), λ is the coefficient and A is the surface of the reservoir.

The coefficient a is used to determine the volume of the water storage V in the karst underground, located towards the spring or in the catchment of the river section between two measurement profiles, by using the following expression:

$$V = 86400 \, (Q_0/a), \tag{3.10}$$

where V is expressed in m^3, and Q_0 in m^3 s^{-1}, while time t is expressed in the given units. According to the change of the outflow coefficient a in time, the differences in the outflow regime can be established, as presented in Figure 3.18 A, B. Both greater permeability of the karst massive, i.e. larger karst fissures, and greater effective porosity result in a greater outflow coefficient. These conclusions appear to be logical, at first sight, but unfortunately, they are accurate only under certain very restricted conditions, which are frequently not satisfied. Formula (3.9) has been developed for the water outflow from the porous medium (i.e. laminar flow) and assuming that the area of the reservoir A from which the water flows is constant (see Fig. 3.18 C) expression (3.9) can be obtained according to expressions (3.11 a) and (3.11 b):

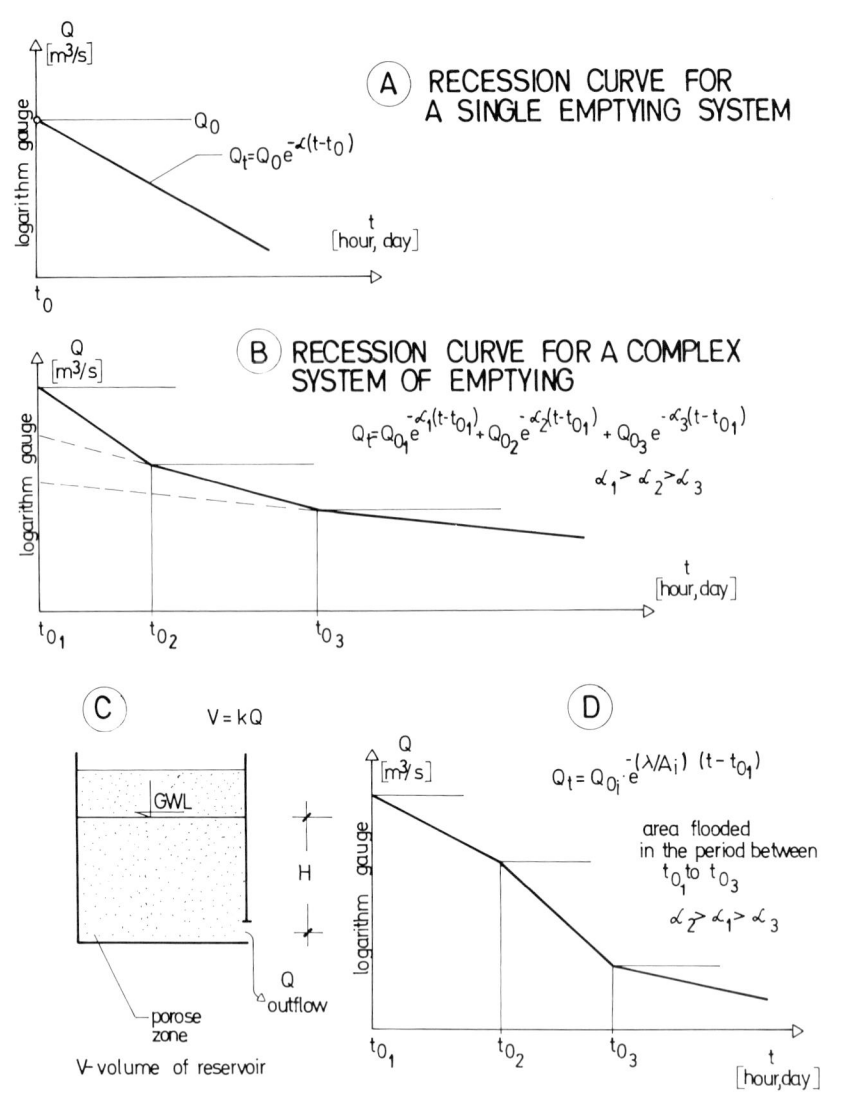

Fig. 3.18 A–D. Analysis of the recession curve for a single emptying system (**A**) and for a complex system of emptying (**B**). Concept of linear reservoir (**C**) and adequate recession curves (**D**)

$$Q = -A \, (dH/dt) \, ; \qquad (3.11\,\text{a})$$

$$Q = c \cdot H \, . \qquad (3.11\,\text{b})$$

Expression (3.11 a) represents the equation of the linear reservoir with a constant area A, whereas expression (3.11 b) introduces a constant c, which refers to seepage. By deriving expression (3.11 b) and introducing it into *Eq.* (3.11 a) we obtain expression (3.9) which, consequently represents a formula for the outflow from the porous medium, assuming the area and filtration to be constant. Since most frequently this assumption is not true, this expression has to be applied in practice very carefully, particularly in karst terrains, where filtration and laminar flow are expected to occur. There is another essential drawback of this approach. In karst terrains, especially during flood periods in the poljes of upper horizons, from whose catchments there are inflows to the analyzed spring, an important factor is the significant change of the reservoir area A in time. This change strongly affects the change in the outflow coefficient a. This results in situations presented in Figure 3.18D when coefficient a does not decrease successively in time. Consequently, when applying this method, although more adequate than others, its results should be taken as approximate and they should not be accepted as reliable values. In order to obtain the effective porosity n_e from the recession curves it is necessary, in the analyzed period of recession duration, to define all the net hydrologic inflows in the catchment, and to know the exact or at least approximately the precise volume of the karstified rock mass out of which the water flows. Thus, a greater accuracy in defining the effective porosity can be obtained, since it is frequently exceptionally difficult to define even the karst catchment area, and even more so the depth or the volume of the rocks in the karstified massive.

Consequently, the recession curve should not be used alone to draw conclusions on the effective porosity of the karst mass or on the volume of water in the underground storages. This method can be used for preliminary analyses, and in combination with measurements of both the groundwater level, the infiltration coefficient and permeability. Drogue (1972) postulated the possibility of using the recession curves and the value of the data obtained from it. He statistically analyzed 100 recession curves of the karst springs in France and tried to find an empirical expression best suited to them. He concluded that it was the following expression:

$$Q = Q_0 (1+at)^{-3/2}, \qquad (3.12)$$

where parameter a has the same role as in expression (3.9) (outflow coefficient), but should not be used for defining the water storage in the massive or other more general conclusions. It was essentially a statistical analysis of the measurement data and the choice of the most appropriate empirical function.

The following two expressions for the recession curves have been defined for the Klokun Spring (Yugoslavia):

$$Q_t = 0.980\, e^{-0.055(t-15)}; \qquad (3.13\text{a})$$

$$Q_t = 0.429\, e^{-0.034(t-30)}, \qquad (3.13\text{b})$$

where the time t was measured in days from the top of the hydrograph, i.e. from the beginning of its decrease, and the discharge Q was expressed in $m^3 s^{-1}$. The given recession curves were defined using a standard method based on a large sample including all measurements carried out from 1965 to 1983. The water volume stored in a karst aquifer from the 15th day until $t = \infty$ was $1.97 \times 10^6\, m^3$. The exponential nature of the suggested recession curves suggests that the flow in the component of the aquifer occurs according to Darcy's law.

The third method used for determining, generally speaking, the karst spring capacity is based on the traditional hydrologic approach, on the water budget method for the area under consideration. This method defines, using the known inflows and outflows, the change in the groundwater storage during the hydrologic year. In order to apply this method, the precise boundaries of a hydrologic-hydrogeologic catchment should be known, and all the relevant outflows and inflows into it should be controlled. These conditions can rarely be satisfied in practice, however, hydrologic data on analogous catchments as well as other hydrologic methods can be used in order to obtain results applicable in engineering practice. These budgets should be developed for longer, continuous time periods, i.e. longer than 30 years. This method is used to directly obtain data on the water volume oscillations in underground retention. The determination of the effective porosity is closely related to the data on changes in the groundwater level in time, karst geology (distribution of karstified rocks in space) in the studied region and on hydraulic characteristics of the groundwater flow. The storage capacity and the water circulation in the karst fissures can be perfectly illustrated by experience in closing a temporary spring, i.e. the Obod Estavelle located in the Fatničko Polje in Herzegovina (Yugoslavia). The spring was closed in order to reduce the inflow into the polje and thus to protect the polje from flooding. It was considered that all the water, not flowing underground to the Fatničko Polje, would directly flow to the poljes of lower horizons. After the polje was closed, an intensive rainfall filled the karst massive with water up to 120 m above the spring level. On the closed spring several built-in manometers measured a pressure of 12 atm. Springs appeared at high levels, groundwater destroyed the long-distance power lines and slidings were formed on the road high above (50 m) the Obod Estavelle. Its closing decreased the inflow into the Fatničko Polje from $60\,m^3$ to $12\,m^3\,s^{-1}$. Accordingly, on the one hand, our objective was achieved, i.e. the polje was protected from the flood, but on the other hand, heavy damages were caused on the horizons high above the poljes where groundwater appeared. This damage called for urgent blowing up of the concrete slog used for closing the spring. The previous natural conditions had to be established as soon as possible. As such sudden reaction was not expected, adequate measurements of the groundwater levels could not be carried out. Thus, they could not be used to draw more detailed conclusions on the quantitative characteristics of the karst voids related to the water storage. This example showed in a drastic manner that in this case the storage capacity of the karst medium was not large, although it referred to the bare, well-developed karst of Herzegovina.

The fourth method used for the determination of effective porosity n_e is based on the application of the Theis equation of the logarithmic approximation for non-stationary flow during the pumping process. When applying this procedure, one should bear in mind that only approximate values can be obtained for the karst area. The complex features of outflow in karst are difficult to identify as it is difficult to determine even such basic hydraulic parameters such as permeability and transmissivity. Pumping experiments can most frequently be used to obtain parameters related only to a narrow area surrounding the considered location. The Theis method is applicable to homogeneous media; thus, it can be applied to karst only where such conditions are at least approximately acceptable. Torbarov (1975) believes that in certain karst zones, particularly around the Trebišnjica Spring (Yugoslavia), the karst terrain can be approximated

Table 3.4 Review of effective porosity n_e coefficients on various locations in the world

Ordinal No.	River or region (country)	Author (year)	n_e [%]
1.	Hesbaye (Belgium)	Manjoie (1984)	3–4
2.	Lamalou Spring (France)	Paloc and Thiery (1984)	2
3.	Brévine Valley (Switzerland)	Burger and Pasquier (1984)	0.2–0.4
4.	Limestone ridge between the Dabar and Fatničko Poljes (Yugoslavia)	Milanović (1983)	6–10
5.	Mendip Hills-Somerset (Great Britain)	Atkinson (1977)	0.92
6.	Area around the storage Salakovac on the Neretva (Yugoslavia)	Ivanković and Komatina (1976)	2–3
7.	The Trebišnjica Spring (Yugoslavia)	Torbarov (1976)	1–3
8.	The Trebišnjica Catchment (Yugoslavia)	Torbarov (1976)	1.2–1.5
9.	Buško Blato (Yugoslavia)	Borelli and Pavlin (1967)	1
10.	The Cetina Spring (Yugoslavia)	Stepinac (1979)	0.17
11.	The Grab and Ruda Springs (Yugoslavia)	Stepinac (1979)	0.29
12.	The Rječina Spring (Yugoslavia)	Stepinac (1979)	0.23
13.	Zrmanja on the section Žegar-Jankovića Buk (Yugoslavia)	Stepinac (1979)	0.31
14.	The Ombla Spring (Yugoslavia)	Milanović (1981)	1.4–3.5
15.	The Karst Spring (Libanon)	Mijatović (1968)	3.2
16.	The Upper Zeta (Yugoslavia)	Vlahović (1972)	0.79
17.	The Lower Zeta (Yugoslavia)	Vlahović (1972)	0.79
18.	Pregrada between the Nikšićko Polje and Lower Zeta (Yugoslavia)	Vlahović (1972)	6.07
19.	The central part of the Tennessee Catchment (USA)	Moore, Burchet, Bingham (1969)	0.4–3.4
20.	Lower part of the Cetina Catchment near the Prančević Reservoir (Yugoslavia)	Bonacci (1981)	2.5

as homogeneous. This method was used to determine the effective porosity by Torbarov (1975), and Ivanković and Komatina (1976). The pumping with changing capacity under conditions of non-stationary flow was replaced by natural outflow conditions, and thus the analysis period was restricted to the period with no rainfall inflows into the underground, i.e. the recession period (Torbarov 1976). The effective porosity n_e is computed using the following expression:

$$n_e = 2.25 \cdot T \cdot t_0/x^2, \tag{3.14}$$

where T is the transmissivity coefficient expressed in m^2 s^{-1}, t_0 is the time when the decrease of groundwater in the piezometer ΔR is equal to zero, while x expresses the distance from the observed piezometer to the spring. Expression (3.15) is used to define the transmissivity coefficient T:

$$T = c_1 \cdot \bar{Q}/c, \tag{3.15}$$

where c_1 is the constant, being 0.183 for the radial flow with an angle of 360° (for the spring the angle is considerably smaller, and the constant is somewhat greater), Q is the average discharge in the analyzed period, whereas c is defined by the expression:

$$c = \frac{\Delta R_2 - \Delta R_1}{\log t_2 - \log t_1}, \tag{3.16}$$

where ΔR_i is the variation of the groundwater storage expressed in mm of the groundwater column height in period t_i. Consequently, this procedure can be applied only after the above mentioned conditions have been satisfied, at least partly. The indispensable measurement data include the discharge measurements at the spring (or at the inflow or outflow profile of the karst streamflow) and measurements of the groundwater levels on piezometers surrounding it. Conclusions should not be drawn according to data on changes in the groundwater level on a small number of piezometer or on only one since local conditions of the analyzed measurements point can significantly affect the accuracy of the computation.

In order to obtain an insight into the order of magnitudes, Table 3.4 presents a review of the values for effective porosity n_e in karst. The presented results were taken from the literature without any particular selection. In each case the effective porosity n_e was determined in various ways, and in most publications the method used to determine it is not stated. Understandably, this makes it impossible either to compare results or to make a critical review or a deeper analysis. These data, however, even considering great differences in the methodology used to obtain them, indicate that the karst void system, in general, has no great capacity for water storage. The average values of n_e range from 1 to 2%, and the strongly cracked upper zone, as well as some other parts particularly subjected to karstification and tectonic processes, reach an effective porosity ranging from 5% to a maximum of 10%. These sections, because of their location, and infrequent occurrences, are not essential for the more general regional analyses of karst water resources. These locations have a negative effect and present serious engineering problems in ensuring impermeability of artificial reservoirs in karst.

4 Karst Springs

4.1 General Concept and Classification

Karst springs represent a natural exit for the groundwater to the surface of the lithosphere through the hydrologically active fissures of the karst mass. The springs in karst appear most frequently in the places of contact between the carbonate masses and the impermeable layers (e.g. flysch). The water flows to the surface through the permeable rocks, which are practically insoluble, and sometimes non-karstified (Bögli 1980). Lehman (1932) mentions the karst-hydrologic contrast expressed by the presence of numerous places through which the water sinks into the karstified mass, whereas there are relatively few karst springs. In the early phase of karstification fewer parts of the aquifer are oriented towards one spring, i.e. in this phase there are a great number of springs with a small catchment area. As the hydrologic activity increases the respective catchment area of a spring becomes larger and deeper. Consequently, certain springs stop functioning, and the remaining active springs become larger and have a greater capacity (Bögli 1980). It is already evident in this phase that the catchment area of karst springs changes in time depending upon the water quantity in the aquifer. This variability can be greater or smaller depending upon the local and geologic conditions. This statement can be confirmed by the fact that the water which is underground in one place often emerges in numerous springs distantly located one from another, and its appearance depends on the water quantity in the catchment expressed by the groundwater level. Two characteristic examples of Dinaric karst (Yugoslavia) are shown in Figures 4.1 and 4.2. Figure 4.1 shows a situation with three springs of the Krka and Krčić Rivers, whose water joins the spring water via a waterfall 40 m high, thus forming the Krka River. Dyeing tests performed at eight boreholes have shown that the dye appears only at two springs (Pavičić 1981), whereas the dye from two boreholes did not appear at any of the three analyzed springs. Figure 4.2 presents the case of a spring in the Žrnovnica Bay at the Adriatic sea coast. There are eleven springs on the coast quite near the sea and three submarine springs (vruljes). The capacity of all the surface springs in the dry period is $0.5 \text{ m}^3 \text{ s}^{-1}$, whereas the maximum capacity is $3 \text{ m}^3 \text{ s}^{-1}$. The only spring with permanent fresh water is Spring 1, Spring 3 has the greatest capacity. In dry periods, Springs 2 to 11 discharge salty water with a salinity of 3000 to 11000 mg l^{-1} Cl. In a period of heavy rains the salinity of the water decreases in all springs; it ranges from 50 mg l^{-1} at certain springs to 3000 mg l^{-1} at others. The appearance of this spring zone is conditioned by the boundaries of permeable limestone layers, and impermeable dolomites and flysch stretching along the seacoast. The flysch belt, together with the dolomites, directs

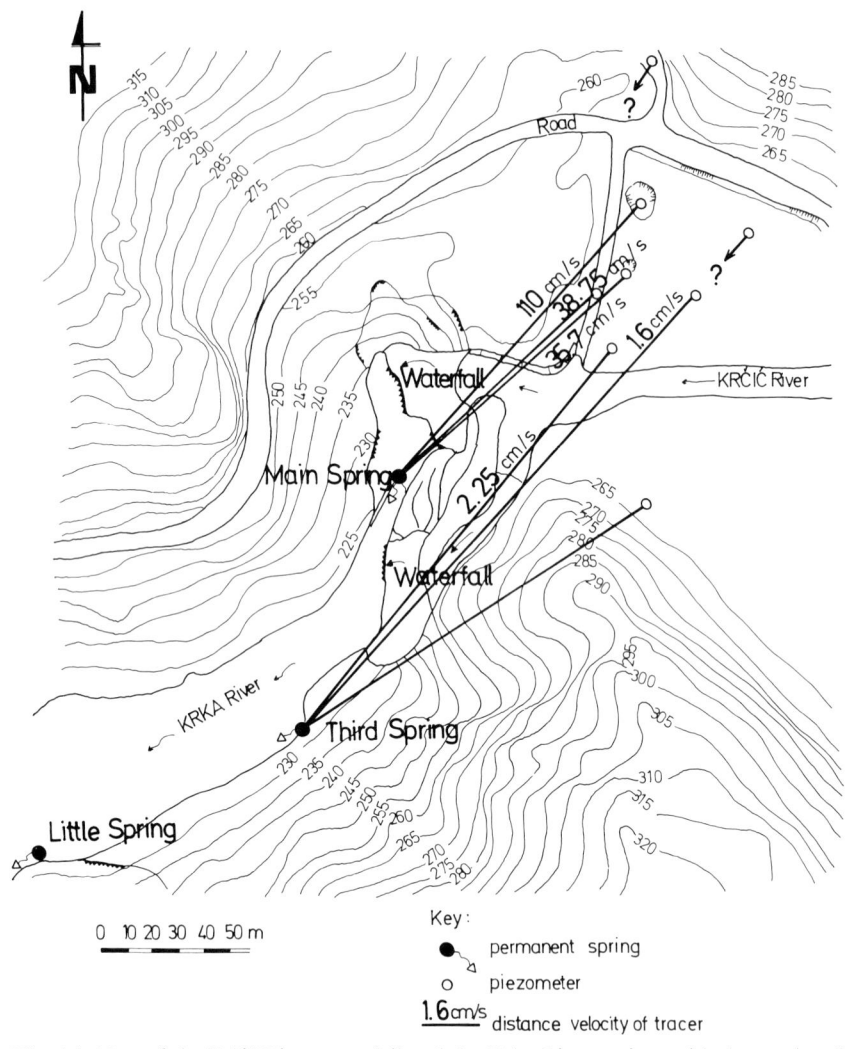

Fig. 4.1. Map of the Krčič River waterfall and the Krka River springs with the results of tracer experiments (Yugoslavia). (Pavičić 1981)

the karst groundwater to the surface directly into the Žrnovnica Bay. Krznar et al. (1970) have tried to define the main groundwater flows from the hinterland to the springs and have, therefore, performed dyeing tests in the piezometers bored in the hinterland. Figure 4.2 shows the established links between the piezometers and springs with the given travelling time of the dye from the piezometers to the springs.

It is quite difficult to precisely classify karst springs. Even in carefully performed tests there are always certain exceptions which deny or at least make the classification uncertain. Bögli (1980) presents three types of spring classifications

General Concept and Classification 51

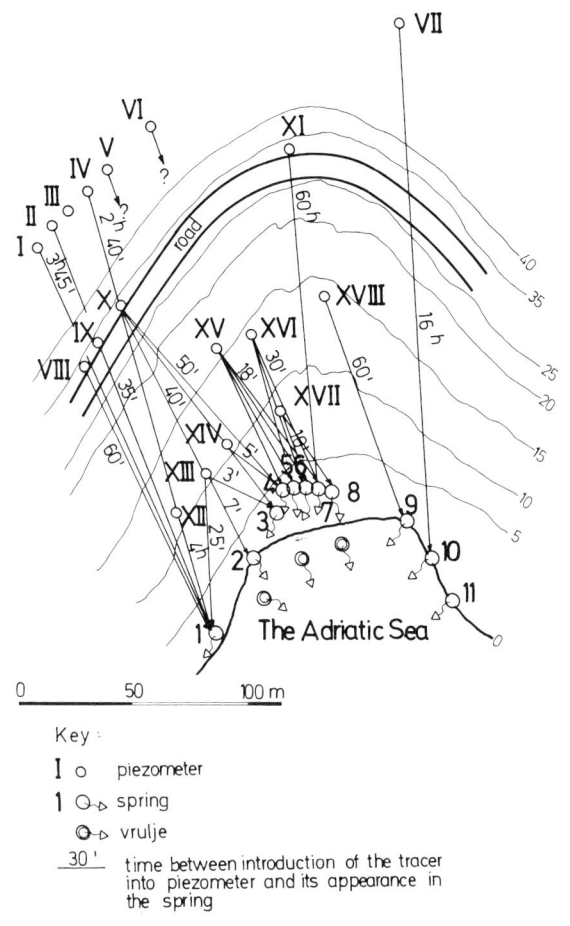

Fig. 4.2. Tracer experiments carried out in the hinterland of the Žrnovica Springs (Yugoslavia). (Krznar et al. 1970)

according to the following features: (1) outflow hydrograph; (2) geologic and tectonic conditions; (3) water origin.

According to the outflow hydrograph, Bögli (1980) distinguishes the following types of springs: (1) perennial; (2) periodic; (3) rhythmically flowing, intermittent or ebb and flow springs; (4) episodical.

According to the geologic and tectonic conditions (Bögli 1980), springs can be divided into: (1) bedding springs (Fig. 4.3 A and B); (2) springs emerging from fractures (Fig. 4.3 C); (3) overflow types of springs (Fig. 4.3 D); (4) ascending springs (Fig. 4.3 E). According to this classification springs may be descending or ascending. Figure 4.3 gives examples of descending and ascending springs.

Taking into account the origin of water appearing at springs, Bögli (1980) suggests the following classifications: (1) emergence; (2) resurgence and (3) exsurgence springs. The origin of water cannot be exactly established for the first type of springs. Most frequently those are large permanent karst springs. In the

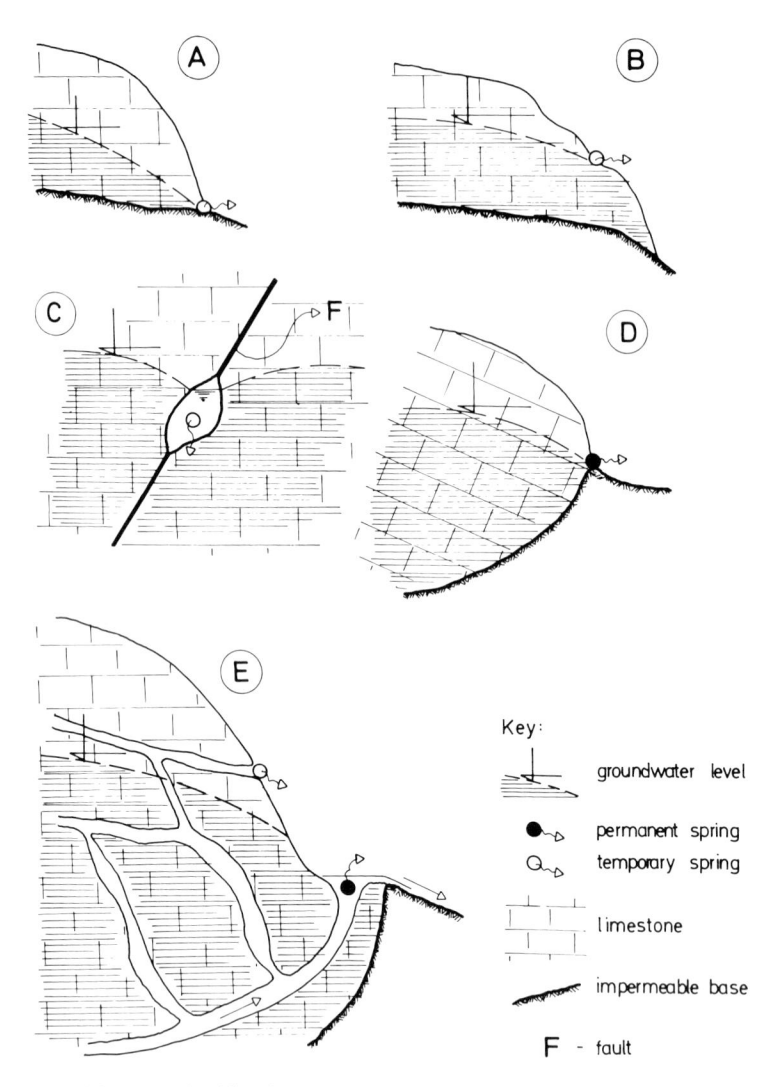

Fig. 4.3 A–E. Classification of karst springs according to geologic and tectonic conditions. (**A**) and (**B**) two types of bedding springs; (**C**) spring emerging from fractures; (**D**) overflow types of springs; (**E**) ascending spring

second case, the groundwater flow reappears at the surface. A characteristic case is presented by the Ljubljanica River (Yugoslavia) which sinks underground near Postojna at the Pivka River. It appears about 9 km from the Planinska Cave as a large flow, the Unica River, enriched by the waters flowing from the eastern part of the karst catchment. The Unica sinks again this time through the numerous ponors in the Planinsko Polje, and it appears at numerous springs of the Ljubljansko Polje about 8 to 10 km from the ponor zone. Figure 4.4A shows the situation of the karst catchment of the Ljubljanica with the calculated hydrologic areas (subcatchments), whereas Figure 4.4B shows some quantitative and

General Concept and Classification 53

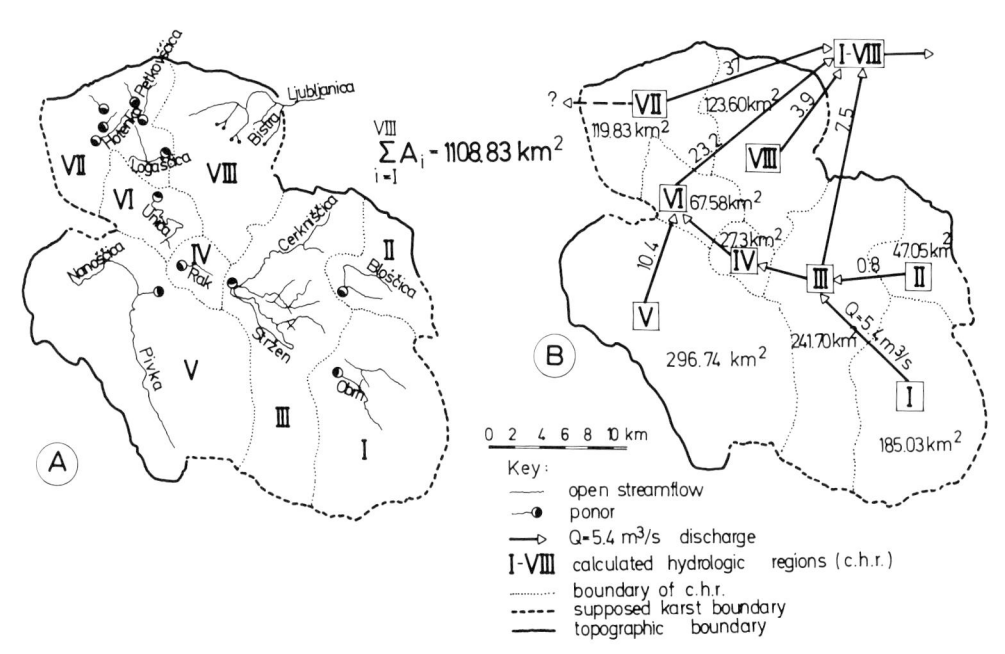

Fig. 4.4 A, B. The Ljubljanica River basin with calculated hydrologic regions (**A**) and hydrologic connections between them (**B**) (1972–1975) (Yugoslavia). (Žibrik et al. 1976)

qualitative hydrologic relationships between the calculated catchment areas (Žibrik et al. 1976). Exsurgence springs exhibit the outflow of exsurgence water of autochthon origin.

Permanent springs are very frequent karst phenomena and the greatest quantity of the groundwater appears at the surface exactly at these places. The groundwater appears in different ways. Large springs often have ascending wells named after the Vaucluse Spring (Fontaine de Vaucluse, Avignon) in France. Figure 4.5 (Magdalenić et al. 1986) shows the location of the permanent karst spring Bulaž in Istra (Yugoslavia). It is a typical ascending karst spring whose capacity ranges from a minimum of 0.1 m^3 s^{-1} to a maximum of 30 m^3 s^{-1}, with an annual average quantity of ca. 2 m^3 s^{-1}. The photograph presented in Figure 4.6 shows another ascending spring situated in the Cetina River Catchment. It is a strong karst spring of the Ruda River. The spring is of the ascending type with an emergence level at 300 m above sea level. The photograph was taken during one of the longest dry periods occurring in that region. The capacity of the spring was quite high even then, i.e. ca. 5 m^3 s^{-1}. The minimum discharge measured in the period between 1974–1984 at the Ruda Spring was 4.42 m^3 s^{-1}, and the maximum discharge was 43.0 m^3 s^{-1}, whereas the mean discharge over several years was 14.03 m^3 s^{-1}. The ratio between the minimum, mean and the maximum discharge is 1:3.2:9.7, which shows that the inflow to the spring has been exceptionally uniform. This spring has a very favourable regime from a hydrologic standpoint, and thus confirms a high regulation capacity of the karst mass in its hinterland.

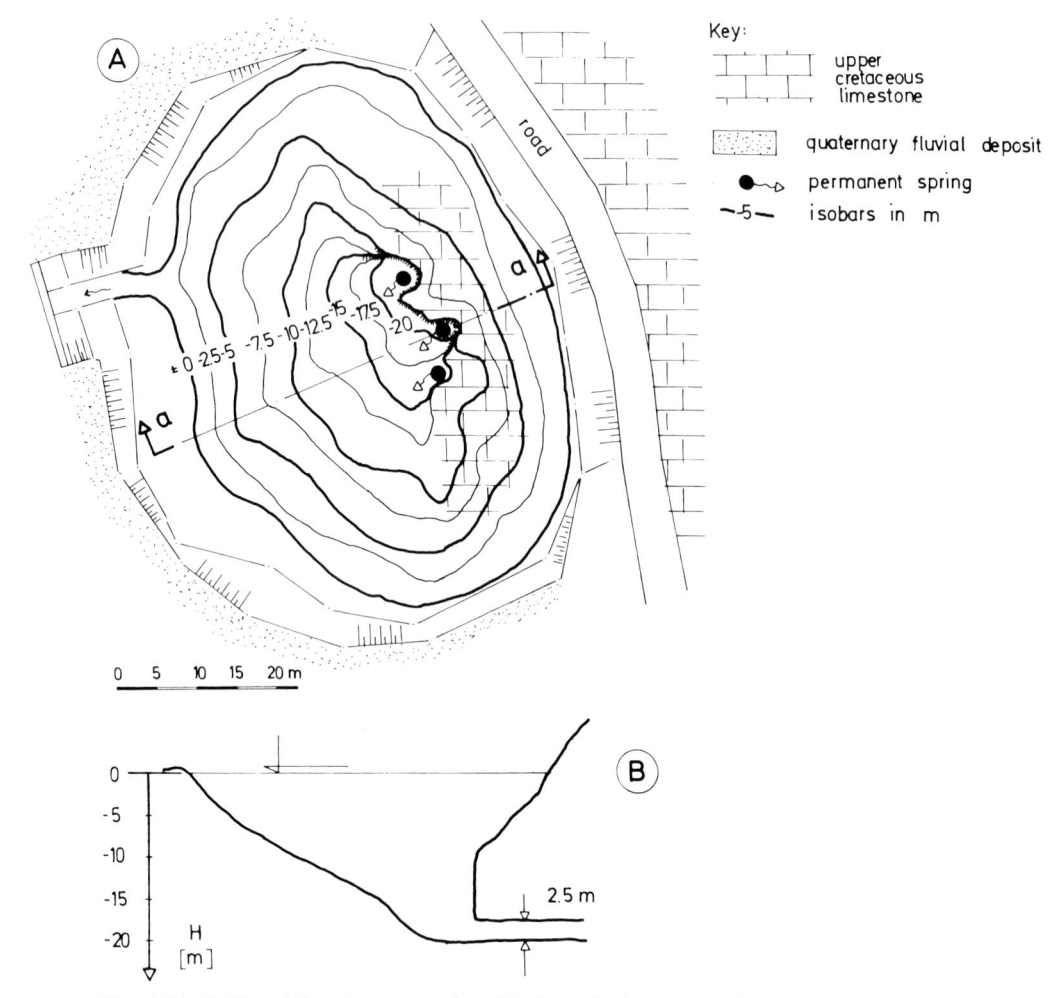

Fig. 4.5 A, B. Plan (**A**) and cross-section (**B**) through the karst spring Bulaž (Yugoslavia)

The photograph included in Figure 4.7 presents a completely different type of spring. It refers to the Studenci spring zone located in the downstream part of the Cetina Catchment. Figure 4.7 presents one of the main springs, but in the very spring zone there are four large permanent springs and numerous smaller permanent and intermittent springs. The springs are located at different levels ranging from 10 to 20 m above sea level, and those levels where water appears at the surface vary according to the condition and level of the groundwater in the hinterland of the karst mass. Therefore, they are called suspended springs. It is practically impossible to carry out individual measurements due to the inaccessible terrain and unfavourable conditions for conducting precise hydrometric measurements. The phenomenon of the Studenci spring zone is caused by the contact between flysch and limestone layers about several hundred meters long. The permeable limestone is situated above the impermeable flysch. According to the

General Concept and Classification

Fig. 4.6. The Ruda Spring (Yugoslavia) October 10, 1985, after a longlasting dry period $Q \approx 5 \text{ m}^3/\text{s}^{-1}$, (taken by Granić)

few and insufficiently reliable measurements, the minimum discharge of the entire Studenci spring zone was estimated to be ca. $2 \text{ m}^3 \text{ s}^{-1}$, whereas the maximum discharge is estimated to range from $18-20 \text{ m}^3 \text{ s}^{-1}$. Suspended springs have been observed in the Tara River Canyon (Yugoslavia), which are of a periodic type.

The Krka Spring is a specific karst phenomenon which deserves special attention (Bonacci 1985). It has been quite well studied from the geologic and speleologic viewpoint. The Krka River Springs consist of three independent parts: (1) the Main Spring, (2) the Little Spring and (3) the Third Spring (Figs. 4.8 and 4.9). The Main Spring is located in the cave directly under the great waterfall. Although there are no precise measurement data it is thought to supply 80–90% of the total water quantities of all the Krka Springs. The Third Spring has the

Fig. 4.7. The Studenci Spring zone in the Cetina River catchment (Yugoslavia), (taken by Granić)

Fig. 4.8. The Krčić Waterfall and the three springs of the Krka River (Yugoslavia), (taken by Granić)

General Concept and Classification 57

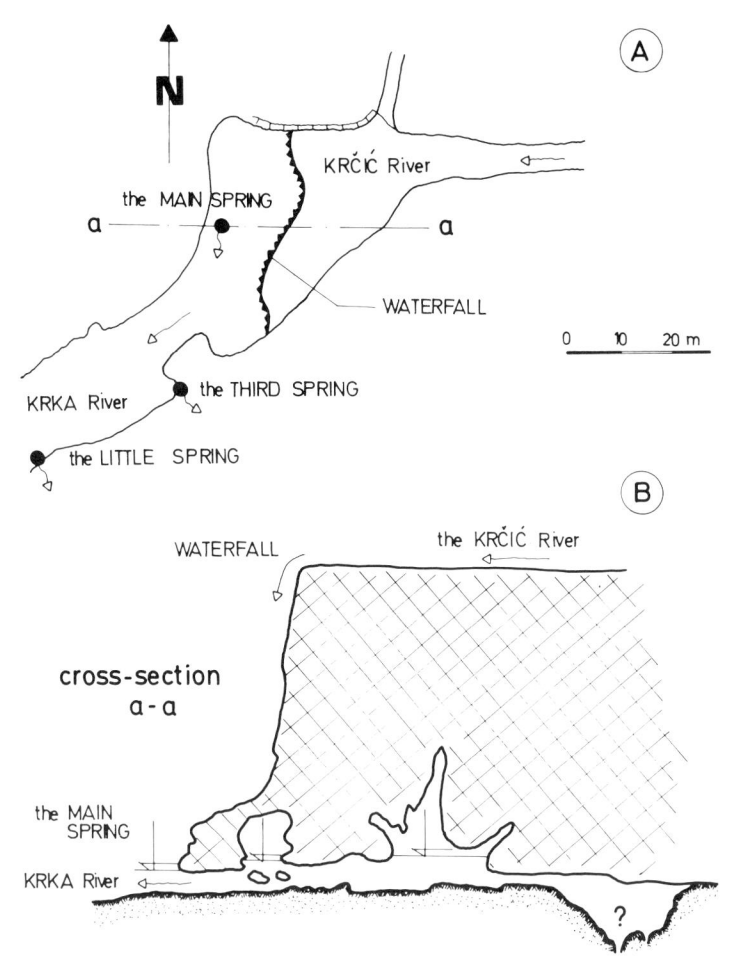

Fig. 4.9 A, B. The Krka River springs map (**A**) and cross-section (**B**) of the speleologically investigated part of the cave with the main spring (Yugoslavia). (Božičević et al. 1983)

smallest discharge; it is situated on the left bank of the Krka 50 m downstream from the waterfall. Figure 4.9 presents the map (A) and the cross-section through the cave (B) downstream from the waterfall. The Little Spring accounts for about 5–15% of all the underground discharge of the Krka River and its capacity is two to five times greater than the capacity of the Third Spring; it is located 150 m downstream from the waterfall on the left bank of the Krka River. The most interesting spring, considering its capacity, is the Main Spring located under the waterfall in a siphon. Speleologists have managed to gain access and take measurements (Fig. 4.9B). Thus, we can conclude that the Krka River is formed partly by the surface flow of the Krčić coming across the waterfall and partly by the above mentioned springs. Its average annual discharge is 12.6 m^3 s^{-1}. The discharge of the Krka Springs is defined simply by subtracting the Krčić discharge measured at the upstream Station 4 from the Krka discharges measured at nearby

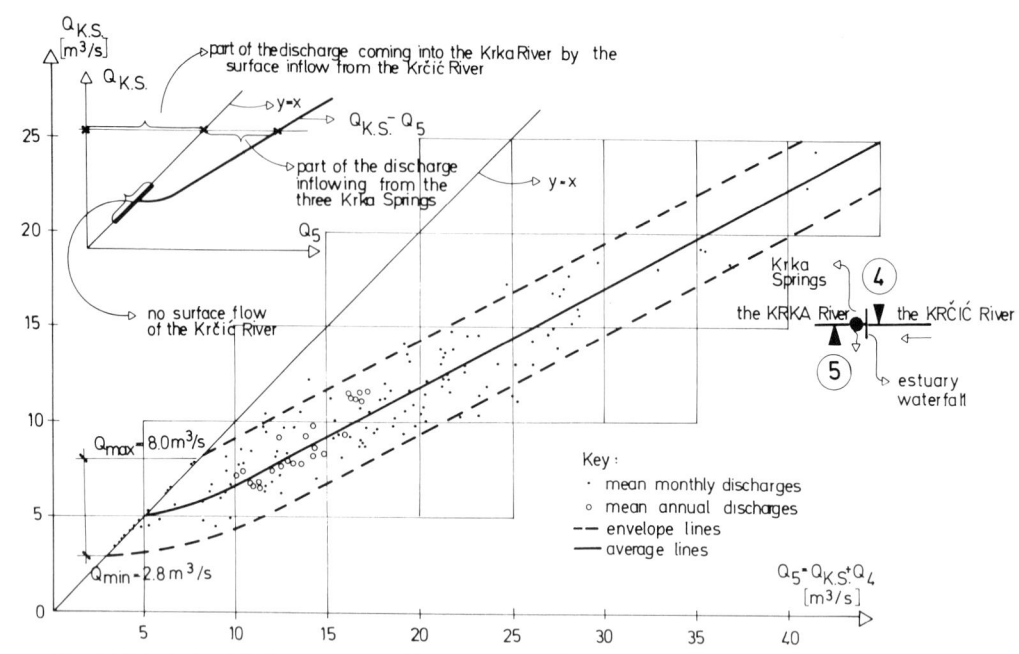

Fig. 4.10. Relationship between monthly and annual discharges of the Krka Springs and Krka Station no. 5 (Yugoslavia)

downstream Station 5 (Fig. 4.10). It amounts to 7.5 m³ s⁻¹ over an average of several years. Figure 4.10 presents the relationship between the average monthly and yearly discharges of the Krka Springs and the total sum of discharges of the Krka Springs and Krčić. The scheme in the upper left corner is given as an explanation of the plotted diagrams. The area between the ordinate axis (the Krka Springs discharge) and the line $y = x$ represents the part of the discharge inflowing into the Krka River exclusively from underground, whereas the second half of the quadrant represents the part of the discharge supplied by surface flow from the Krčić. The points (plotted) on the line $y = x$ represent the situations when the discharge flows exclusively from underground, i.e. when it dries up and there is no flow in the Krčić riverbed. It can be seen that when there is no surface flow of the Krčić, the discharge of the Krka River varies from 2.8 to 8 m³ s⁻¹.

Intermittent springs are most frequently situated in the zone of vertical circulation. The groundwater appears at the surface at these places only in the periods immediately following heavy precipitation. Their hydrographs are most often steep and short. Figure 4.11 presents a typical previously described spring in the Vrgorsko Polje (Yugoslavia). It serves for the discharge of water from the Rastok Polje situated at the horizon about 40 m higher. The spring is active for only several periods each lasting 10 days a year. This type of periodic spring is frequent in the Dinaric karst, but it does not discharge significant hydrologic quantities and is rarely exploited. A completely different type of a periodic spring is the Krčić River Spring, presented in the photograph in Figure 4.12. The photograph was taken on 26 September 1982, when the spring dried up. It is evi-

General Concept and Classification 59

Fig. 4.11. Periodic spring in the Vrgorsko Polje. December 28, 1982, $Q \approx 0.6 \, \text{m}^3/\text{s}^{-1}$ (Yugoslavia), (taken by Granić)

Fig. 4.12. The Krčić River spring area (Yugoslavia) September 26, 1982, $Q = \emptyset$, (taken by Granić)

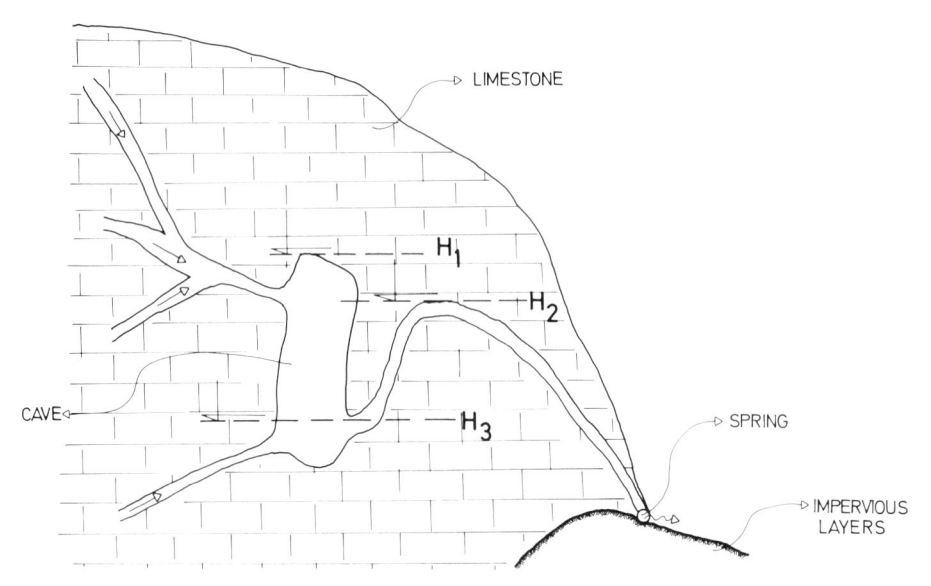

Fig. 4.13. Schema of rhythmic spring functioning

dent that there is no covering soil and, hence, no conditions for vegetation either in the entire spring zone or in the upstream part of the catchment. The subcutaneous zone is well developed, and, thus, the water quickly sinks into the vadose zone through this area. The average discharge of the Krčić between 1980 to 1984 was 5.64 m^3 s^{-1}, and the maximum measured discharge was 22.5 m^3 s^{-1}. In the mentioned period the spring was dry an average of 46 days a year. In 1984, it was dry for only 12 days, and in 1983, the dry period lasted for 94 days. At the Krčić Spring the groundwater appears at the surface from the alluvial deposits carried there by strong and short-lasting currents. The spring level varies approximately 10 m depending upon the groundwater level.

Rhythmic springs or potajnice are an interesting karst phenomenon. They occur relatively rarely and do not play an important role either from the hydrologic or economic standpoint. Gavrilović (1967) reported that only 30 to 40 such springs exist in the entire world; five are located in Yugoslavia and function during the whole year as rhythmic springs. The scheme of their functioning is presented in Figure 4.13. When the water level in the cave rises to H_2, all water in the cave from H_2 to H_3 suddenly outflows. This emptying is effected according to the siphon principle. When the water level is above H_1, and when the inflow through the underground channel and the cracks in the karst mass are great, i.e. greater than the maximum capacity of the siphon, then the spring does not operate rhythmically, but functions as any other karst overflow spring.

Estavelles belong to a special group of springs together with the springs appearing under the water surface (either fresh or sea water). Estavelles have a double hydrologic function. In one period they operate as ponors. This happens most frequently in the dry period of the year, when the groundwater levels are low, and then they are situated under the outflow openings. Very often caves take over the

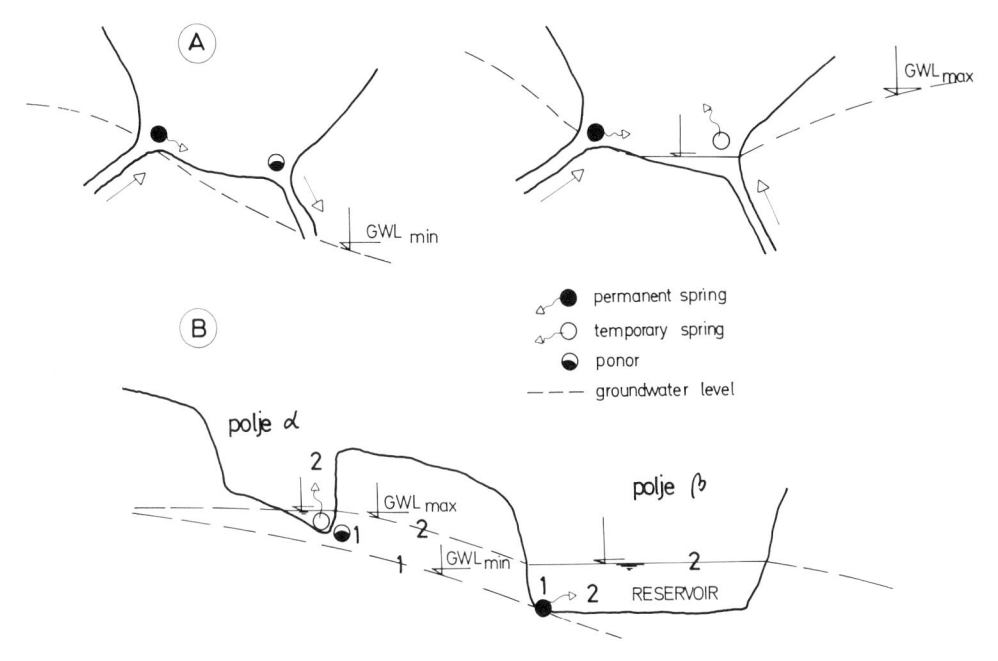

Fig. 4.14 A, B. Explanation of the estavelle functioning as ponor and temporary spring (**A**). Formation of estavelle in situation when the lower horizons are flooded (**B**)

function of estavelles. In the wet period of the year estavelles function as springs. Estavelles most frequently appear in the middle part of the large poljes, between the spring zones located in the upper part of the polje and the springs zones located in the lowest part of the polje. Estavelles can be situated in the lowest part of the polje; in the case of smaller poljes, they are located as cascades separating larger poljes. Figure 4.14 presents the principles of the estavelles functioning partly as ponors and springs; their functioning depends on the groundwater level. When the polje is flooded, estavelles can temporarily function as submergence springs, and when the groundwater level is lowered they become ponors, i.e. take over the function of water evacuation. Estavelles have been discovered in the riverbeds of certain karst rivers in the Dinaric karst (Cetina, Krčić). Dukić (1984) mentions a special type of estavelle, i.e. Gornje-poljski vir in the Nikšićko Polje (Montenegro, Yugoslavia). During the greatest part of the year, it has the form of a lake with a diameter of 85 m and with an average depth of 30 m. In the warm dry period of the year (most often from May to September) it functions as a ponor with the inflow from the Sušica River with a discharge of 100 to 500 $l\,s^{-1}$. At the end of the dry period, after the first heavy autumn rains, the water in the lake becomes turbulent and completely disappears in the ponor after a short time. After 20 to 50 min have passed the muddy water reappears from underground, fills up the shortly emptied lake and flows into the Nikšićko Polje down the Sušica riverbed. The same cycle is repeated regularly every year.

Sometimes ponors situated in the poljes of upper horizons in certain periods act as springs (Fig. 4.14 B): this happens when the poljes of lower horizons are

Fig. 4.15. Vruljes in the Žrnovica Bay (Yugoslavia), (taken by Granić)

flooded or after the construction of storages in karst, i.e. due to the work of man. This phenomenon occurs when the water levels in the artificial storage are so high that they dictate, as boundary conditions, the raising of the groundwater levels in the hinterland of the karst mass. Milanović (1986) described one of those cases; it occurred during the construction of the Bileća Storage in Herzegovina (Yugoslavia).

The existence and functioning of estavelles is proof of the fact that the catchment areas in karst change with time depending on the groundwater levels in the karst mass.

Another phenomenon in karst is the springs below the water level, referring either to the fresh water of the rivers, natural lakes or artificial storages, or the salty sea water. In the latter case, the submarine springs are called vrulje. The photograph of vruljes given in Figure 4.15 refers to the Žrnovnica Bay (Yugoslavia), whose location is represented in Figure 4.2. Those submarine springs, as all other large vruljes, are visible by the circles appearing on the sea surface. When the submarine spring is deep, or has a small capacity there are no visible circles.

Sublacustrine springs often appear in limestone Alpine Valley lakes (Switzerland, Austria, Italy and France). Numerous springs have also been discovered at the bottom of the Skadar Lake (Yugoslavia and Albania). During the periods of heavy precipitation the springs emerge also from the fissures located at the water surface of the lake. Bögli (1980) reported three possibilities of their origin, considering the aeration of the spring after the formation of the lake, i.e. underground erosion and the development of morphologic and/or tectonic processes of the valley and lake bottom.

Submarine springs (vruljes) emerge from small and large karst joints. Very frequently vruljes emerge from submarine caves, and their openings are found at the bottom of the inundated sinkholes (doline). The existence of submarine springs is caused by the lowering of the karst erosion basis in the Pleistocene period. In the most recent geologic past, i.e. after the last deglaciation, the sea level was lower than at present. Šegota (1968) reported that ca. 25 000 years ago, the level of the Adriatic Sea was 96.4 m lower than at present. Then, it represented the erosion basis for the karst area, so that the springs appeared at that level. As the sea level was raised the springs came under the water level, and thus new channels were simultaneously formed in the karst, together with the new, higher positions of the coastal springs. This process is still going on, and Šegota (1968) predicted that it will last for 1900 years, and that the sea level will be raised another 1.17 m. This complex process involves another phenomenon, i.e. the fact that the coast of the Adriatic Sea has been gradually lowered, and is still being lowered. The situation is either similar or identical with other seas. It should be observed that these actual processes will influence for a long time the relations and the position of the submarine and coastal springs in karst. The change in the position of the springs along the coast depends upon the local geologic structure of the coastal belt. The existence of a great number of channels situated at different levels, through which the water circulates below and above the sea, results in the mixing of the fresh and saline sea water. The water in the vruljes, however, as well as in the coastal springs, is most frequently brackish, with its salinity increasing in the dry and decreasing in the wet period, when the discharges of the coastal springs and vruljes are more significant.

It should be noted that there are some saline springs deep inland. Fritz (1978) mentions the Boljkovac Spring in the vicinity of Zadar (Yugoslavia) located at a level of 3.25 m and 2.5 km distant from the Adriatic seacoast. The spring exhibits a very specific characteristic, i.e. its water is saltier when its discharge is great than when it is small. Fritz (1978) explains this phenomenon by the connection of the spring with the sea water. Great discharges are supposed to activate the existing siphon connections. It should be emphasized, however, that this phenomenon has not been entirely explained. The problem related to the process of salinization of coastal springs will be discussed in Section 4.5.

Vruljes have been discovered worldwide: in the Persian Gulf; along the coast of the USA, in the vicinity of New York, Florida and California; along the coasts of Cuba, Mexico, Jamaica, Chile, Hawaii, Australia and Japan; and in the Black Sea. They are most frequent in the area of the Mediterranean basin. Those with the greatest capacities are found along the coasts of Libya, Israel, Libanon, Syria, Greece, France, Spain, Italy and Yugoslavia. Fifty large vruljes have been observed along the eastern coast of the Adriatic Sea in Yugoslavia. About 80% of all vruljes are situated at the sea bottom, up to 10 m, and only a few are located at greater depths, to a maximum of 50 m. Figure 4.16 presents a scheme of the sea bottom with two vruljes, together with its geologic structure.

The variation in the discharge capacity through the vruljes varies significantly during the year. Only a small number of vruljes function permanently as springs.

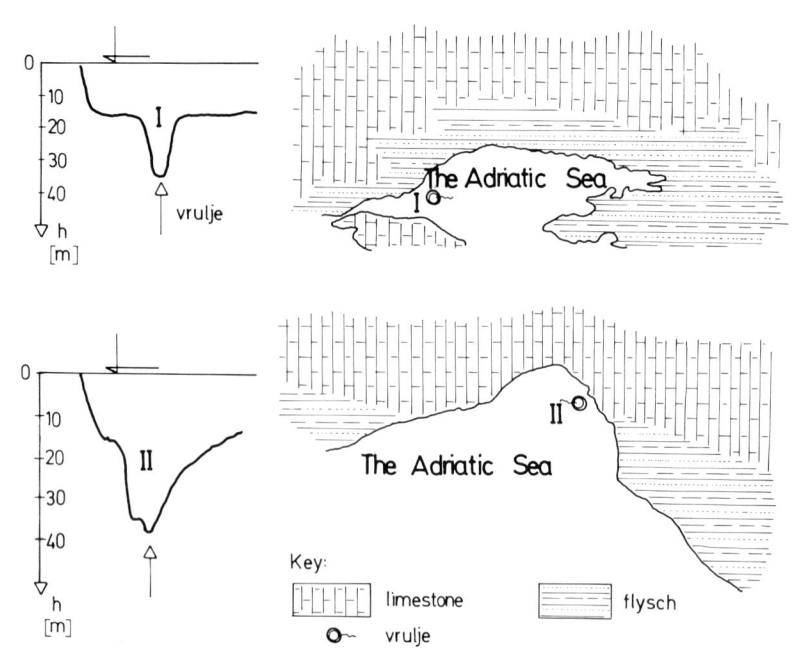

Fig. 4.16. Geologic structure and cross-section through two submarine springs in the Adriatic Sea (Yugoslavia). (Alfirević 1970)

Most dry up in the warm and dry period of the year, and some take over the function of ponors, but only for a short time.

In the Adriatic Sea area (Yugoslavia) very frequently there are a great number of karst spring phenomena located in a relatively narrow area (not greater than 1–5 ha), i.e. surface coastal springs of fresh water, brackish coastal springs and vruljes. The water quantity in the coastal springs with fresh water varies depending upon the tidal flow. Figure 4.2 presents one of those cases in the Žrnovnica Bay (Yugoslavia).

Alfirević (1966) analyzed the vruljes of the Kaštela Bay from morphologic, geologic, sedimentary, hydrogeologic, geotectonic and hydrologic (temperature and salinity) standpoints. Submarine measurements have showed that morphologically these vruljes correspond to the inundated fossile dolines from the pre-alluvial age formed in the continental phase of the Dinaric coastal karst. The salinity and temperature of the vruljes are directly influenced by the precipitation in the catchment area. Precipitation directly results in the circulation of the groundwater which appears in the vruljes and in the entire coastal springs system. Alfirević (1966) established the existence of the homothermis of water at the location of the vruljes generally in autumn and winter, when the average air temperatures range from 0 to 4 °C. Homothermic features of the sea water in the vicinity of the vruljes have not been observed in the same period, since the water was warmer at the bottom and colder at the surface. The salinity of the water also varies according to depth, with lower salinity at the bottom and greater at the surface. The situation is inverse with sea water.

In the summer when there is a small quantity of water in the vruljes (they often dry up) the temperature range, according to depth, is identical in the sea water and in the water above the vruljes. The temperature is lower at the bottom and higher at the surface. The situation is the same with the salinity range, which proves the decrease in the activity of the vruljes or their complete disappearance. Generally, in summer the salinity is greater than it is in winter.

Since vruljes frequently are only one element of a larger spring system, the organization of the hydrologic measurements is a very complex task since these measurements should encompass a wider area. The easiest and most frequent measurements refer to the water temperature and salinity and their distribution along the vertical. It is presently impossible to carry out precise measurements of the water quantity outflowing from the vruljes. It should be emphasized that the standard methods of tracing groundwater by dyeing tests have not yielded satisfactory results.

Considering the hydrologic, and in particular the civil-engineering standpoint, i.e. the water exploitation for practical purposes, the most interesting springs are those present at the beginning of strong, open streamflows. In Yugoslavia, such springs are called "river heads". Taking into account the surface forms of the river heads they can be (Petrović 1983): (1) cave springs; (2) hidden springs; (3) fissure springs; (4) spring eyes (lakes) and (5) spring systems.

Cave springs represent direct outflows of water from the karst underground through the cave whose opening is visible. They are quite numerous worldwide. Karanjac and Günay (1980) reported that the Dumanli Spring (Turkey) is the largest karstic spring in the world issuing from one single orifice. Its minimum discharge after a very long dry period is about 25 $m^3 s^{-1}$, and its maximum discharge is estimated to be over 70 $m^3 s^{-1}$. After the flooding of the Cetinjsko Polje (Yugoslavia) which occurred on 19 and 20 February 1986, the discharge from the Obod Cave was estimated to be greater than 100 $m^3 s^{-1}$. The Obod Cave has an opening 23 m in diameter and 14 m high and is the "orifice" of the Crnojevića River. The catchment area covers 100 to 120 km^2 with an average yearly discharge of ca. 8 $m^3 s^{-1}$. During the previously mentioned 2 days, the precipitation in the spring catchment exceeded 1000 mm, which caused the flooding of the Cetinjsko Polje located in the spring catchment area of the Crnojevića River, which simultaneously influenced the outflow of a great quantity of water from the Obod Cave.

Hidden springs are those springs where the cave openings are closed due to erosion and landslides. Huge blocks of stone and boulders hide the openings of permanent or temporary springs. It is difficult to establish the actual position of the spring in such situations. The water percolates through a layer of stones (frequently alluvial or river deposits) and the level of its appearance at the surface can vary for several tens of meters. At the Žrnovnica Spring near Split (Yugoslavia), the water emerges at an elevation of 60 to 90 m above sea level depending upon the groundwater level in the karst mass. The Crnojevića River emerges from a hidden spring during a dry period, whereas great quantities of water outflow from the Obod Cave, for a short time, only during the exceptionally wet periods when the poljes on the higher horizons, located in the catchment of the Crnojevića River, are flooded.

Fissure springs result from well-developed fissures in the limestone, i.e. due to the existence of large, linked, vertical and sloped fissures (1 to 10 cm large). Essentially, this phenomenon can be explained by the divergence of groundwater flows immediately before their appearance on the surface. The fissure springs may be covered by a fluvial deposit so that they are basically hidden springs. Usually these springs have a low capacity, ranging to a maximum of 5 m^3 s^{-1}.

Spring eyes are often called a lake type of the karst springs. In such cases, the groundwater karst flows emerge at the surface from the cone depressions in a form similar to sinkholes (doline). These springs represent the outflow from the open caves of siphon channels. Figure 4.5 presents a situation and a cross-section through a karst spring, Bulaž in Istria (Yugoslavia), which is a typical example of a spring eye. The spring of the Cetina River (Yugoslavia) consists of a great number of such spring eyes.

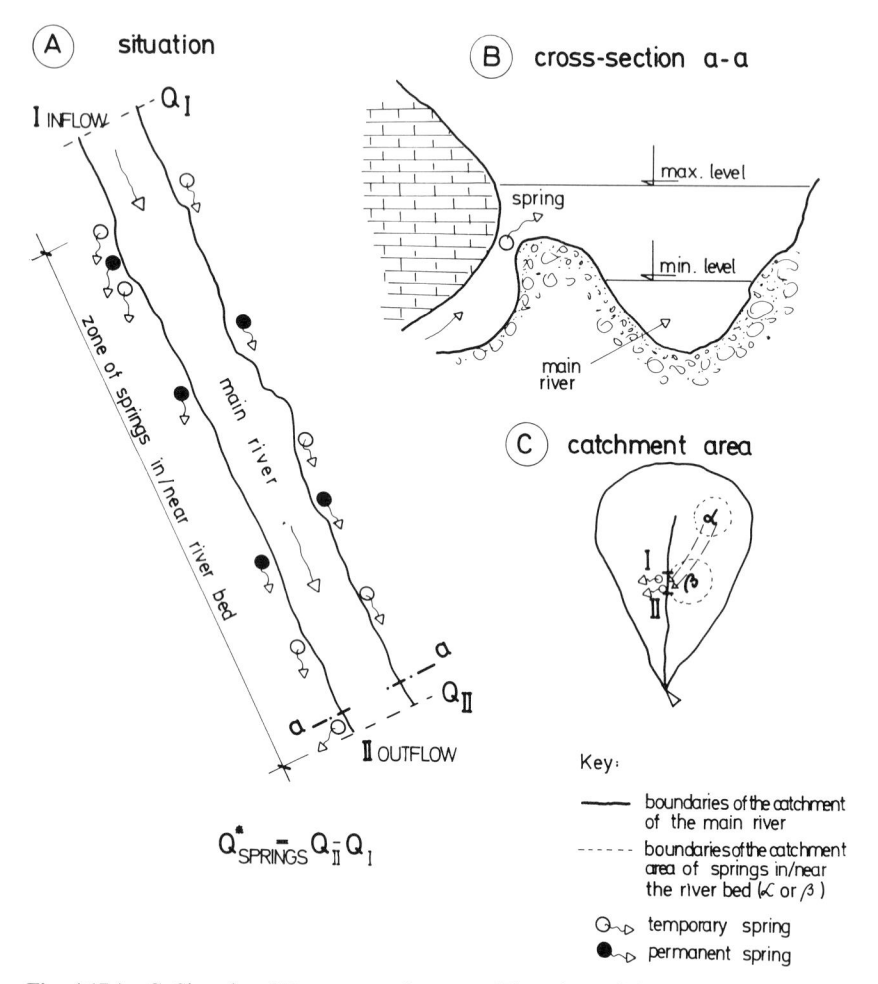

Fig. 4.17A–C. Situation (**A**), cross-section $a-a$ (**B**) and possible catchment areas (**C**) of the karst springs in/near the river bed

Spring systems contain a great number of springs belonging to different or even all types of springs. The phenomenon of spring systems is related to wide karst regions and a well-developed hydrologic system of water circulation. The springs of the Crnojevića River and the Zeta River (Yugoslavia) represent complete spring systems.

A special type of spring is represented by springs in the open riverbed of larger rivers in karst (Fig. 4.17). Such springs are usually numerous and represent an entire spring zone distributed along the length of the river. Most frequently they appear on both banks of the main river as in the case of the Mrzlek Springs on the Soča River (Yugoslavia). Habič (1982) reported that this spring zone is 400 m long, and that the water emerges at 18 places: 9 places on the left and 9 on the right bank. According to small variations in the temperature of these springs, between 8° and 10°C, it has been concluded that the water is retained underground for a long period of time. Karanjac and Altug (1980) refer to a spring zone along the Manavgat River (Turkey), downstream from the Oymapinar Dam. The zone contains 25 medium-size karst springs distributed along both river banks for a length of 1000 m. Some of these springs are permanent and others temporary. The elevation of the water emergence at the springs depends upon the levels of the groundwater flowing to the spring zone. The water level and discharge in the main river are not influenced by the water level and capacity of the spring zone. Consequently, the situations vary from those when the springs are flooded to those when the water emerges at a level significantly higher than the river water level. The longest duration occurs when the water levels in the springs are higher than the water level in the open streamflow.

Very often it is impossible to measure the discharge of each spring; therefore, it is necessary to carry out hydrometric measurements at the inflow (I) and outflow (II) profiles (Fig. 4.17) of the main streamflow. The capacity of the entire spring zone can be calculated by subtracting the downstream discharge from the upstream discharge of the main river course. The existence of the springs or spring zone along the banks of large open karst streamflows points to the complexity of water circulation in karst, and to the problems in exactly defining the catchment areas of the karst springs.

4.2 Discharge Curves

This section refers to the definition of the discharge curves of the springs which depend upon the groundwater levels in their hinterland. They are defined according to the basic hydraulic laws, i.e. it is a graphic and/or analytical definition of the ratio between the discharge and the difference in elevation of the water level influencing and causing that discharge. These investigations and the discharge curves of the karst springs defined in this way significantly help engineers in the hydrogeologic and geologic investigations of the water circulation in karst. It should be emphasized, however, that there is no special hydraulics referring to karst water circulation and the basic hydraulic laws have been applied to the karst situation. A few examples will be presented to illustrate how this method can be applied to reach conclusions on the number, position an dimensions of the systems for water transportation in the karst mass, including the section from the piezometer whose levels were used as an independent variable to the spring itself.

Consequently, piezometric data carry all important information according to which all major conclusions are drawn. This analysis should be carried out extremely carefully, since it has been shown that each piezometer does not yield information on the groundwater levels in the wider area of the karst medium influencing the spring outflow, particularly under the conditions of the circulation in karst. If the piezometer does not satisfy these requirements, it will not be possible to define the spring discharge curves using its data. If the flow is prevalently turbulent and takes place in rough channels under pressure, with significant local losses, the quadratic law of resistance is established. Consequently, the spring discharge depends upon the square root of the difference between the piezometric levels measured at two arbitrary points.

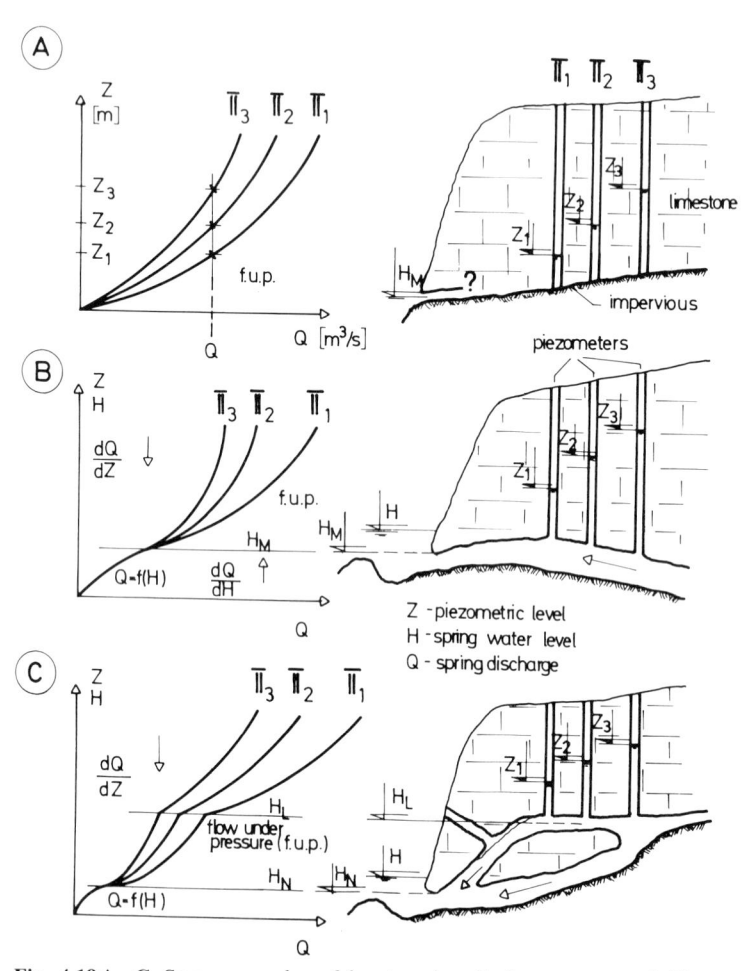

Fig. 4.18 A–C. Some examples of karst spring discharge curves. **A** Flow under pressure for whole water level amplitude; **B** flow with a free water course for $H < H_M$ and under pressure for $H > H_M$; **C** case with two pipes into the karst massive

Discharge Curves

Figure 4.18 gives a schematic presentation of several possible situations. In cases when there is an outflow with a free watercourse, the spring discharge is most frequently the function of the water table in the cross-section along. When the flow is formed under pressure, the spring discharge Q can be defined by the following equation:

$$Q = a\sqrt{2g\,(Z_i - H_N)}. \qquad (4.1)$$

This equation refers to all curves in Figure 4.18A, to the section of the curve for $H > H_M$ in Figure 4.18B and to the section of the curve for $H_N < Z < H_L$ in Figure 4.18C. For the area in Figure 4.18C, when $Z > H_L$, the equation becomes more complex and has the following form:

$$Q = a_1\sqrt{2g\,(Z_i - H_N)} + a_2\sqrt{2g\,(Z_i - H_L)}. \qquad (4.2)$$

Parameter a has the dimension of an area and can be expressed as the product of the mean cross-sectional area of the main inflow channel A and discharge coefficient c. Using this procedure it is possible, in some cases, to determine the mean area of the cross-section of the main inflow channel. The expressions (4.1) and (4.2) for the discharge curves of the springs have been defined according to Hajdin and Ivetić (1978), Hajdin and Avdagić (1982) and Bonacci (1982).

Figure 4.19 presents a possible situation of aquifer formation in the karst conditioned by tectonic processes. The fault caused the bifurcation of the aquifer. The discharge of a permanent karst spring A can be defined only according to the data obtained by piezometers $\mathbb{T}_1$ and $\mathbb{T}_2$, located in the lower aquifer, whereas the groundwater level in the piezometer of the upper aquifer, $\mathbb{T}_3$ and $\mathbb{T}_4$, does not directly influence the capacity of spring A.

The discharge curves for the Krka Spring have been defined according to the previously presented principles and measurements obtained by piezometers (Bonacci 1985). Figure 4.20 presents the discharge curves of the Krka Spring as a function of the groundwater level as observed in the six piezometers located in

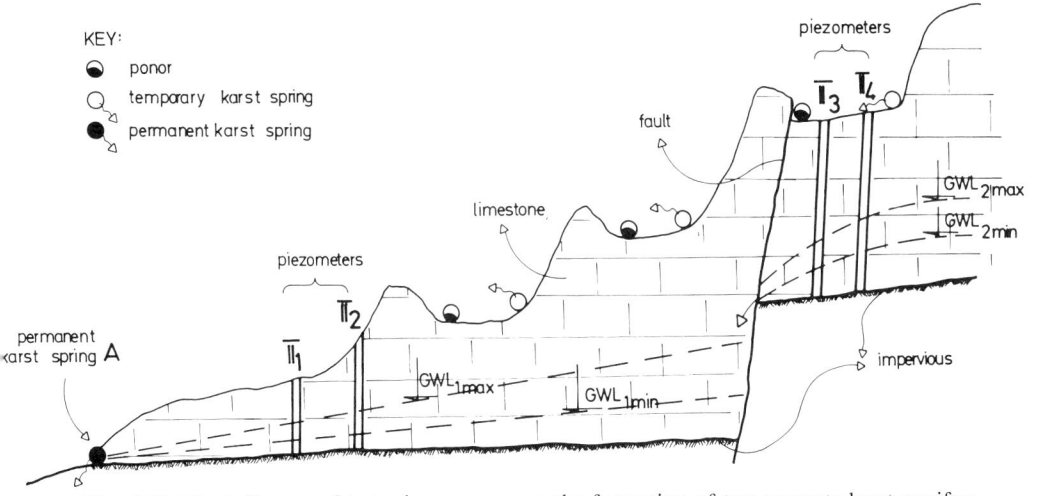

Fig. 4.19. The influence of tectonic processes on the formation of two separate karst aquifers

Karst Springs

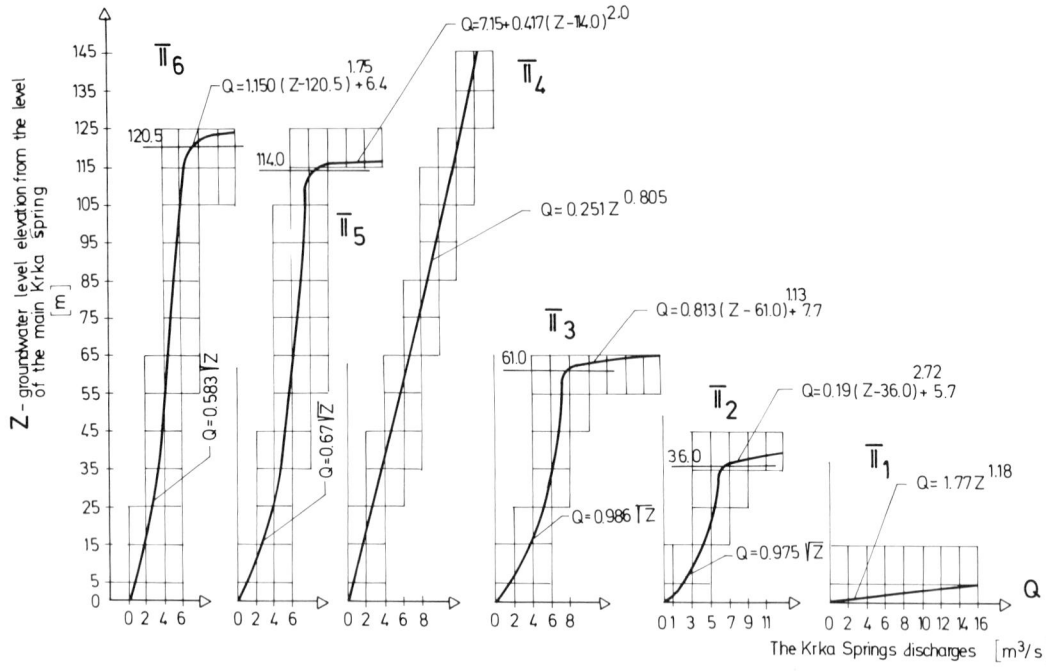

Fig. 4.20. Discharge curves of the Krka Springs (Yugoslavia)

the Krčić Catchment. At five piezometers presented in Figure 4.20 (the exception being piezometer T_4) there is a height up to which the discharge has the same characteristics as the flow under pressure. Beyond this limit the discharge curve is formed according to the free surface flow. In piezometer T_1 the discharge curve does not reach the limit and yet it has the parabolic shape characteristic of a flow with a free surface. Since this piezometer is located near the Krka Spring it does not actually measure flow under pressure. On the other hand, piezometer T_4 is very far away from the open flow of the Krčić (Fig. 4.21) and the limit level does not appear; these piezometers may well be situated outside the zone with free surface flow. It may well be that quite close to the surface flow of the Krčić, there is an underground flow which behaves as an open flow. This flow could take place through pipes, though this is not likely, or through a system of intensely fractured rock due to karstification. All equations for discharge curves were derived using 5-day averages of groundwater levels and the respective discharges of the Krka Springs because groundwater levels in the piezometers are observed only two or three times a week. Figure 4.20 shows that the general expression (4.1) was used between the spring discharge Q, as a dependent variable, and the groundwater level 2 from the level of the Krka Spring in meters, as an independent variable. Parameter a can be defined as:

$$a = Ac, \qquad (4.3)$$

where A represents the hydraulic measurement cross-section and c is the discharge coefficient, always smaller than, but close to l, which ranges in karst from 0.8 to

Discharge Curves

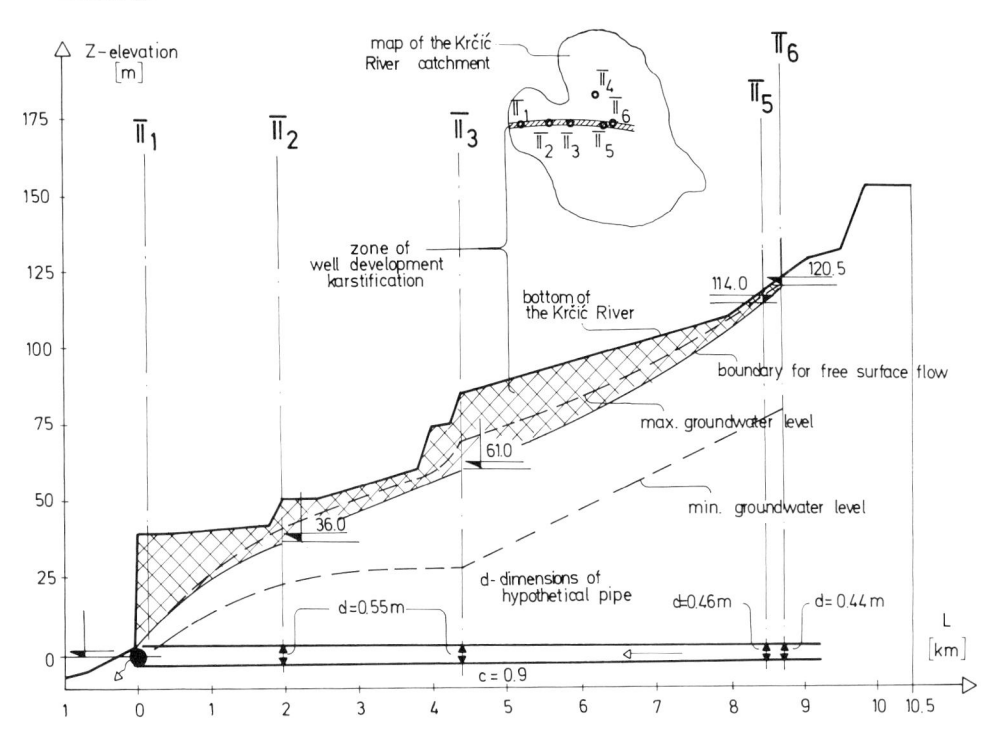

Fig. 4.21. Longitudinal cross-section of the Krčić river-bed with explanation of the conduit flow and diffuse flow toward the Krka Springs in the Krčić Catchment (Yugoslavia)

0.95. Assuming that $c = 0.9$ and that the cross-section is a circular pipe, diameters may be calculated as presented in Figure 4.21. This characterizes the underground water flow in the Krčić Catchment towards the Krka Springs. These characteristics were obtained by analyzing the discharge curves of the Krka Springs. The assumption of the existence of a pipe system and of the dimensions of the pipe is speculative; it is more likely to be a zone with fissures within which the flow can be approximated as flow under pressure. It can be stated, however, with certainty that the water flow to the Krka Spring is affected not only by infiltration, but mostly by turbulent flow in karst "pipes" for which the quadratic law of resistance holds true.

Another example of defining the spring discharge curve in karst as dependent on the groundwater level in the hinterland will be presented on the discharge curve of the Donja Zeta Spring (Vlahović 1972). Figure 4.22 presents the situation of the upper horizons (Nikšićko Polje, Yugoslavia) and the spring in its natural state. The discharge curve shown in Figure 4.23. illustrates the dependence of the Donja Zeta Spring discharge on the function of the groundwater levels obtained by piezometer 142. It should be emphasized that measurements were carried out on 177 piezometers over 6 years, but only the measurements performed on a few piezometers have made it possible to define precisely the relationship between the discharge measured at the Rošca gauging station as a dependent variable and the

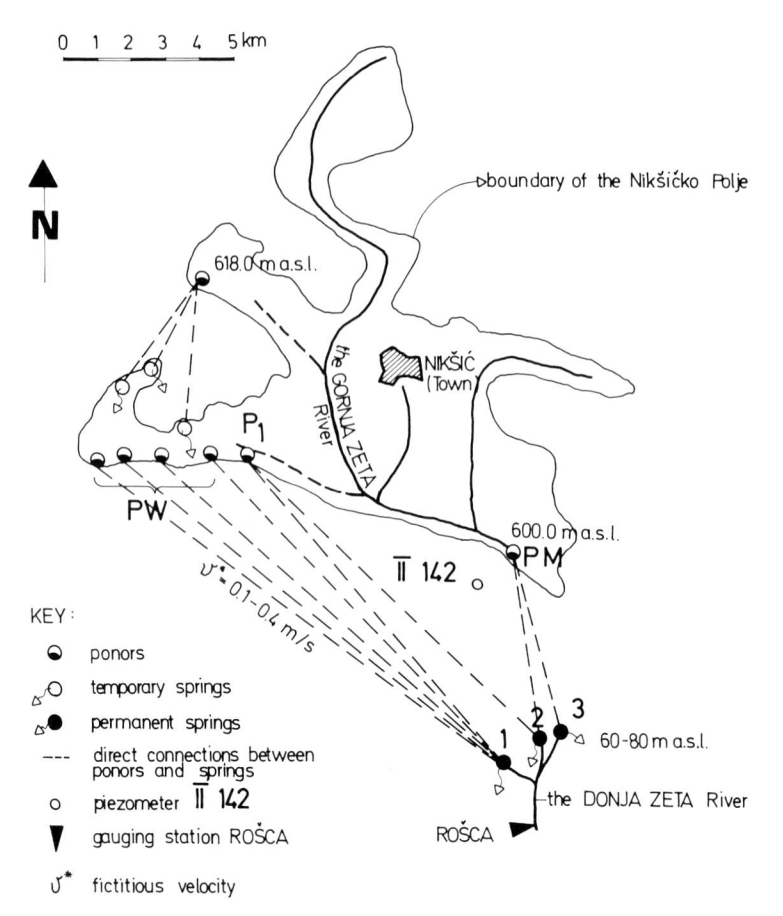

Fig. 4.22. Schematic map of connections between the Nikšičko Polje (600 m a.s.l.) and three springs of the Donja Zeta River (Yugoslavia). (Vlahović 1983)

groundwater level as an independent variable. Therefore, it is necessary to be extremely careful and to carry out extensive investigations in situations in which quantitative hydrologic conclusions have been drawn in this way, i.e. according to the groundwater levels in the hinterland. This procedure can yield significant data, but is quite expensive since it calls for the boring of a great number of piezometers, although data obtained from most of them will not be directly used in defining the discharge curves. The schematic presentation in Figure 4.22 shows that there are a great number of ponors in the Nikšičko Polje, and that they are all emptied via three springs which subsequently form the Donja Zeta River. The minimum level of the Nikšičko Polje is 600 m above sea level, and the springs are located at an elevation of 60 to 80 m. Spring 3 changes its outflow level beyond 100 m above sea level for a short period of time, depending on the groundwater levels. The ponor tracings by dye tests have shown that almost all the water from numerous west ponors PW appears at Spring 1. The dye from the ponor P appears at Springs 1 and 2, whereas the water from the largest ponor PM at the bot-

Discharge Curves

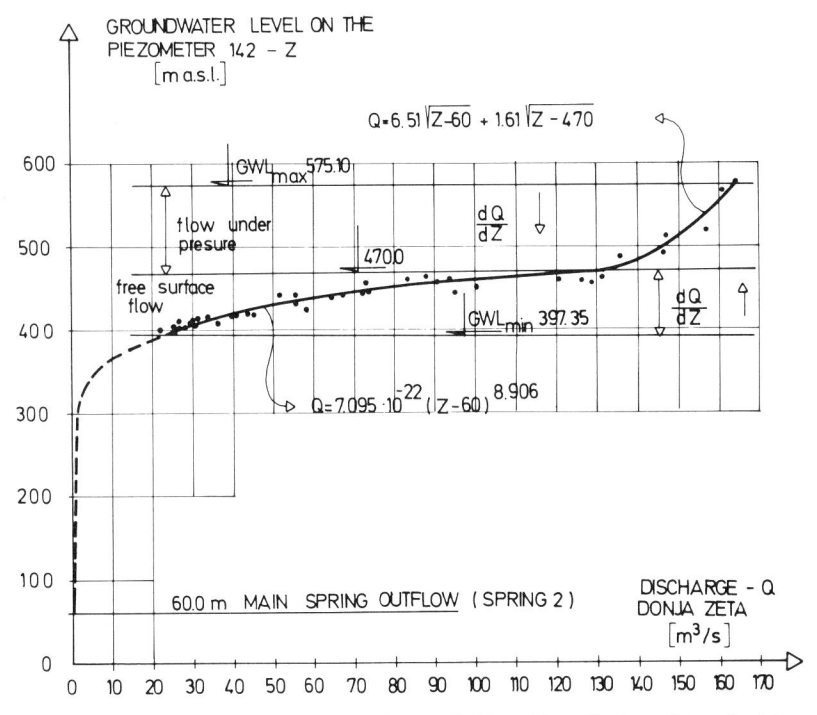

Fig. 4.23. Discharge curves for the Donje Zeta River Springs (Yugoslavia)

tom of the Nikšićko Polje directly appears at Springs 2 and 3. The fictitious velocity at the section west ponors-Spring 1, ranges from 10 to 40 cm s^{-1}. This velocity is great and proves that there is a direct linkage with the karst underground channels and that they are well developed. It should be noted that Spring 1 has the smallest discharge as compared with Springs 2 and 3. Greater fictitious velocities of water circulation can be expected along the sections ponors-Springs 2 and 3. Measurement data on those sections were not available. In this actual case the fictitious velocity is similar to the velocity of the groundwater, since the groundwater channels are expected to be situated at the shortest distance between the ponors and the springs. The discharge curve is defined in Figure 4.23, i.e. from an elevation of 60 m above sea level to an elevation of 460–470 m above sea level. This curve is typical on the flow along the free water surface. Until the water level in the piezometer reaches the level of 470 m above sea level, the flow through the groundwater channels is not under pressure. Beyond that level, the channels are completely filled with water, and the flow comes under pressure. Figure 4.24 presents one possible explanation of the relationship between water circulation and groundwater channels. The discharge curve of the form presented in Figure 4.23 evidently shows that the system is a complete whole, and that the aquifer is linked with it along the entire stretch from the Nikšićko Polje to the Donja Zeta Spring. It is certain that the dimensions of the groundwater channels have the dimensions of 0.3 to 1.5 m^2. The estimations of the dimensions are given in Figure 4.24. There are channels between certain ponors which make it

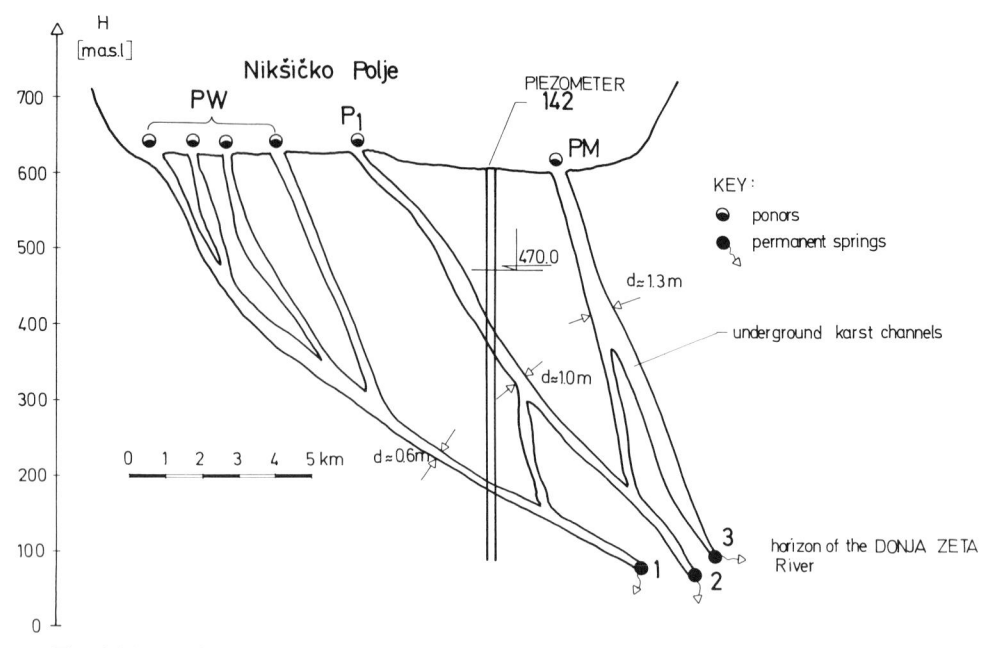

Fig. 4.24. Longitudinal cross-section of the Nikšićko Polje and the Donja Zeta River horizons with hypothetical explanation of the linkages between the ponors and the springs (Yugoslavia)

possible for the water and dye to appear at two springs. The flow is generally turbulent, whereas the laminar flow through the small karst cracks is not significant in this actual case. The situation previously described is very favourable for hydrologic investigations, but does not occur frequently in karst. Practically all of the water from the upper horizon of the Nikšićko Polje appears at a nearby spring in only one lower horizon; furthermore, the karst intercatchment along the section ponor-springs is not too large (covering ca. 5% of the total catchment area) so that it does not significantly contribute to the hydrologic budget, and thus does not influence the formation of the discharge curve of the Donja Zeta Spring. Very frequently in karst the water from one ponor appears at springs located on different horizons, considering the situation and time of their appearance. In such cases, the suggested approach to the definition of the spring discharge curve does not yield satisfactory results. It should be noted that the discharge curves should not be defined according to the simultaneous and instantaneous measurement of the groundwater levels and discharges. It is necessary, however, to use the average values of the data obtained by measurements carried out over 5 days or 1 day depending upon the development of the connections within the karst mass (channels). The use of the average values makes it possible to eliminate from this analysis the momentary situations which do not present a real picture of the general condition, primarily of the groundwater levels and then of the spring discharges. In order to analyze the groundwater levels, it is necessary to collect data covering the entire karst aquifer area, and not only the local groundwater levels in one piezometer. Using the average value it is possible to achieve that objective. In the periods of long-lasting recession, the averaging can be shorter, since all processes are slower in that period. The averaging is necessary particularly in the periods of heavy rainfall and in the period immediately following it, as each piezometer and spring exhibit a different reaction depending on its local characteristics.

4.3 Hydrograph Analysis

The hydrograph of a karst spring has the characteristics of the path along which the rainfall flows from the catchment surface to its natural exit at the surface. The hydrographs of different karst springs differ since the morphology of their underground channels network is different. Knežević (1962) and Knežević and Voinović (1962) studied the influence of the following factors: inflow discharge quantity, development of the underground channels network, the number of positions and the radii of the narrowings on the channels, on the form of the curve showing tracer concentration at the exit (simulated spring). The very simple, natural situations were simulated under laboratory conditions. The concentration of the tracer (dye, salt or radioactive matter) is momentarily injected into the ponor, and the tracer appears at the piezometer or spring as a diffuse wave; its change of shape in time can be simply determined by collecting the water samples. Undoubtedly, the forms of the diffuse waves differ due to the different morphologic characteristics of the groundwater channel network stretching between the locations where the samples are injected and ultimately collected. Knežević and Voinović (1962) carried out such laboratory investigations to obtain, for the different schemes of the channel network, the respective schematized form of the diffuse waves at the springs. Thus, they tried to reconstruct the morphology of the karst hinterland according to the measurements carried out at the springs. Figure 4.25 presents the results of three experiments conducted in order to study the influence of the branches of the main watercourse channel on the form of the diffuse wave. The variety of underground karst forms cannot be easily reproduced by the models, since there are significant quantities of water in the larger channels flowing to the spring by a laminar flow through a system of karst cracks as well as those by turbulent flow. Both types of flow occur simultaneously; the turbulent flow is dominant in the periods immediately following heavy precipitation, whereas the laminar flow becomes dominant in dry periods, especially if they are long-lasting and it forms the recession part of the hydrograph of the karst spring. All the previously mentioned factors influence the form of the hydrograph of the karst spring. Numerous interdisciplinary investigations should be carried out, beyond the field of hydrology, in order to discover the morphologic structure of the karst mass by studying the spring hydrographs. This book emphasizes the hydrologic-methodological approach and does not discuss the other procedures, although they are highly recommended.

The analysis of a karst spring hydrograph is a typical hydrologic approach which can be used to draw general conclusions on the basic structure and morphology of the karst mass elements which are significant for the transportation of water to the spring under consideration (Back and Hanshaw 1965). This can

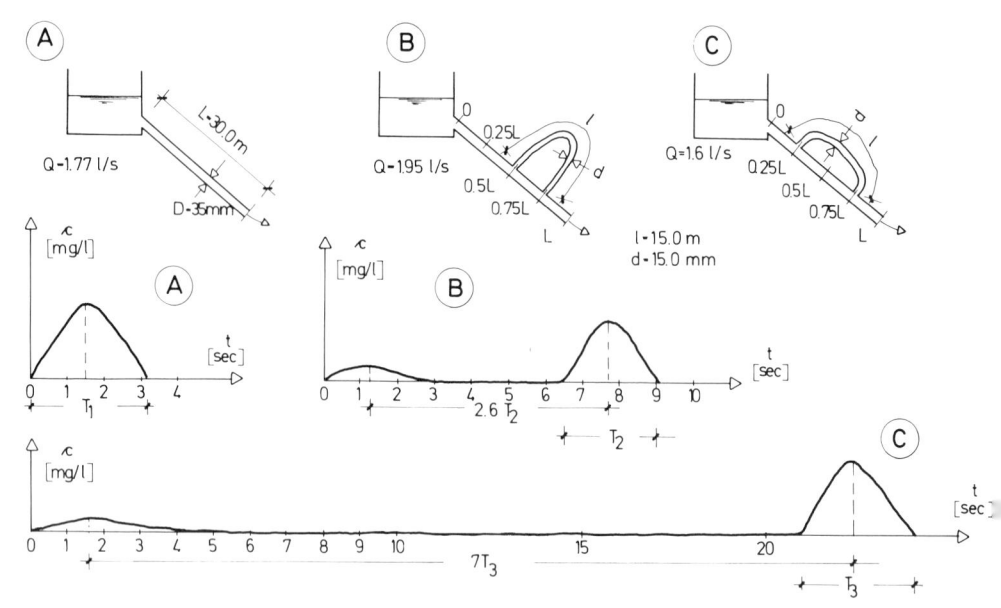

Fig. 4.25 A–C. Three laboratory experiments (**A**); (**B**); (**C**) of the changes in tracer concentration c in dependence on the underground karst channels position. (Knežević 1962)

be illustrated by Figure 4.26 taken from Newson (1973). Four springs reacted to the same rainfall with different hydrographs, and with different hardness. The water of the conduit flow is typically much less hard because of its rapid and concentrated flow through the limestone than the water of the diffuse flow (Newson 1973). Spring D is a typical spring fed mostly by the conduit flow, whereas Spring A is prevalently filled by diffuse or percolation water. The other two types of springs are a combination of the two, so that the conduit flow is dominant in Spring C, but there are probably two dominant channels along which the water flows to Spring C by a turbulent flow. In the case of Spring B, there is a significant time lag between the occurrence of the hydrograph peak and the cessation of the rainfall. The hardness is slightly decreased, which proves the fact that the water has been retained in the karst underground for a long time. Since the hydrograph peak is quite high and occurs with a significant lag after precipitation, the existence of the karst morphology in the hinterland of Spring B could be given by the existence of a large underground storage area, probably a cave, filled by water which percolated after the analyzed rainfall. The filling of the underground storage by water leads to the raising of the groundwater level, thus causing an increase in the spring discharge. It can be supposed with certainty that the underground storage space is located relatively distant from the spring since it takes a long time to establish the piezometric level from the storage area to the spring which subsequently causes the increase in the spring discharge.

In certain situations the pulse or pressure wave plays a dominant role in the formation of the spring hydrographs. Precipitation results in the formation of the pulse wave which takes the water to the spring. This water has been sinking

Hydrograph Analysis

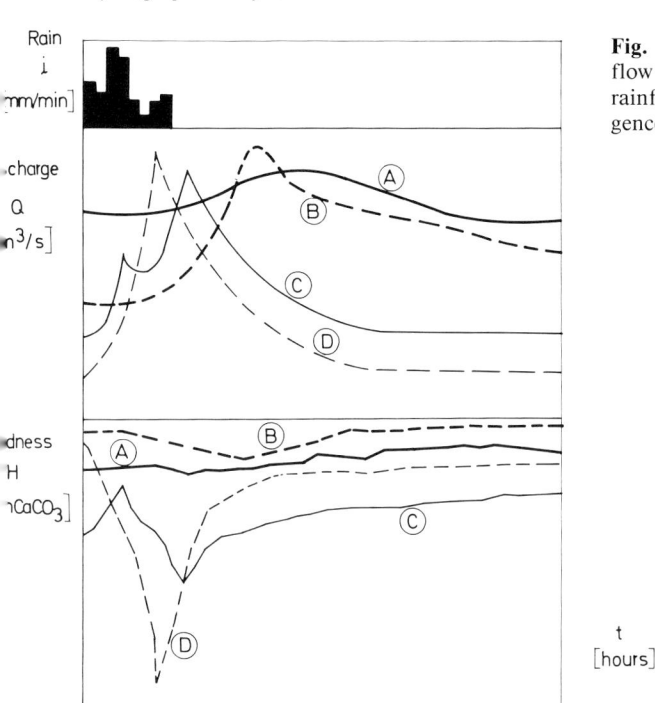

Fig. 4.26. The different response of flow and hardness values to heavy rainfall, shown by a group of resurgences. (Newson 1973)

underground long before the occurrence of the rainfall. According to Yevjevich (1983), water is retained underground for several years in the karst regions of central Turkey. In the Mediterranean basin springs the water appears immediately after rainfall in the central part and is carried exclusively by the pulse wave. This phenomenon can be favourable considering the environmental protection against pollution, since the contaminating substances are retained underground for a long time and there is a real possibility of its chemical desintegration. In some cases, however, its effects can be contrary to the above described situation. The velocity of the pulse wave is 50 to 100 times greater than the normal velocity of water circulation in karst. If the average velocity of water circulation in karst is supposed to be from 1 to 5 cm s^{-1}, then the velocity of the pulse wave is from 5 to $10 = \text{m s}^{-1}$. This value is obtained according to the following estimations. The maximum velocity of the elastic wave through a rigid pipe is ca. 1000 m s^{-1}, whereas its velocity through a deformable pipe is ca. 500 m s^{-1}. The channels through which the water percolated in karst are elastic and deformable due to the numerous karst cracks, and accordingly, the velocity of the pulse wave is estimated to be from 5 to 10 m s^{-1}. Understandably, in certain situations this velocity can be greater or smaller.

The application of the theory and practice of unit and synthetic flood hydrographs to the flow in karst is possible only after considering some specific features. These features refer primarily to the differences in the flow velocity and, consequently, the formation of the hydrograph forms, i.e. the conduit and diffuse flow. Since synthetic flood hydrographs can be set against recorded flood hydrographs and can be defined as hydrographs computed from basin characteristics or basin parameters and rainfall data under the conditions of different types of karst, special consideration should be given to the selection of the easily measurable basin characteristics. In addition to the usual characteristics such as area, slope, length of the longest channel, etc., for defining the synthetic flood hydrographs in karst, some other geologic and hydrologic characteristics of the catchment should be considered: the area of the catchment covered by vegetation, the possibility of overland flow formation, etc. Synthetic hydrographs based on basin parameters are generated by conceptual and mathematical models. In the conceptual models, physical operators of the basin lake storage effect and translation action are grouped together to simulate basin behaviour. In the mathematical models, mathematical functions are used to transform rainfall into a runoff hydrograph.

A conceptual scheme for classifying carbonate aquifers in terms of groundwater flow systems and hydrogeologic settings proposed by White (1969) was revised and extended in 1977, to include factors of relief, structure and areal extent of the aquifer. The structural and topographic settings (White 1977) act mainly by determining the arrangement of karstic rocks with respect of sources of recharge and points of discharge. Lithologic and stratigraphic factors control the degree to which the conduit permeability is developed. White (1977) shows that a distinction between diffuse flow aquifer systems and conduit flow systems can be made in terms of their response to transient recharge events. According to the author's experience in defining the synthetic hydrographs in the Dinaric karst, it is possible to recommend the conceptual models based upon the application of the linear reservoir. One condition which must be completely satisfied is that the boundaries of the catchment area should be exactly defined and determined. Very often this condition cannot be satisfied because of the water circulation in karst; this fact will be discussed in the next chapter (see Sect. 4.4).

The recession hydrograph analysis was previously discussed in Section 3.2. It should be added that the application of the Maillet expression (Eq. 3.9) is very frequently used in practice, but it is often applied in the wrong way and often not carefully considered. The volume of the water stored at the beginning of recession (Eq. 3.10) is determined according to the parameter from Eq. (3.9). Investigators often select only one hydrograph (referring to the longest dry period) and, accordingly, draw conclusions on the water quantity stored in the karst underground. Coefficient a (the storage depletion factor) is defined by applying the method of the least squares, using the data on the smallest discharges. These data are, however, often unreliable (due to the easily explained factors), and hence, coefficient a cannot be considered as a stable value. Therefore, a detailed hydrologic and statistical analysis of all the available recession parts of the hydrograph should be carried out in order to define this factor as precisely as possible. The volume of water stored at the beginning of the recession obtained by the storage depletion factor a, using expression (3.9), refers only to the statical reserves of the groundwater.

An analysis of the spring hydrographs can be used to define the volumes of closely connected small and large fissures of the karst massive from which the water flows into the springs. These fissures consequently result in effective porosi-

Hydrograph Analysis

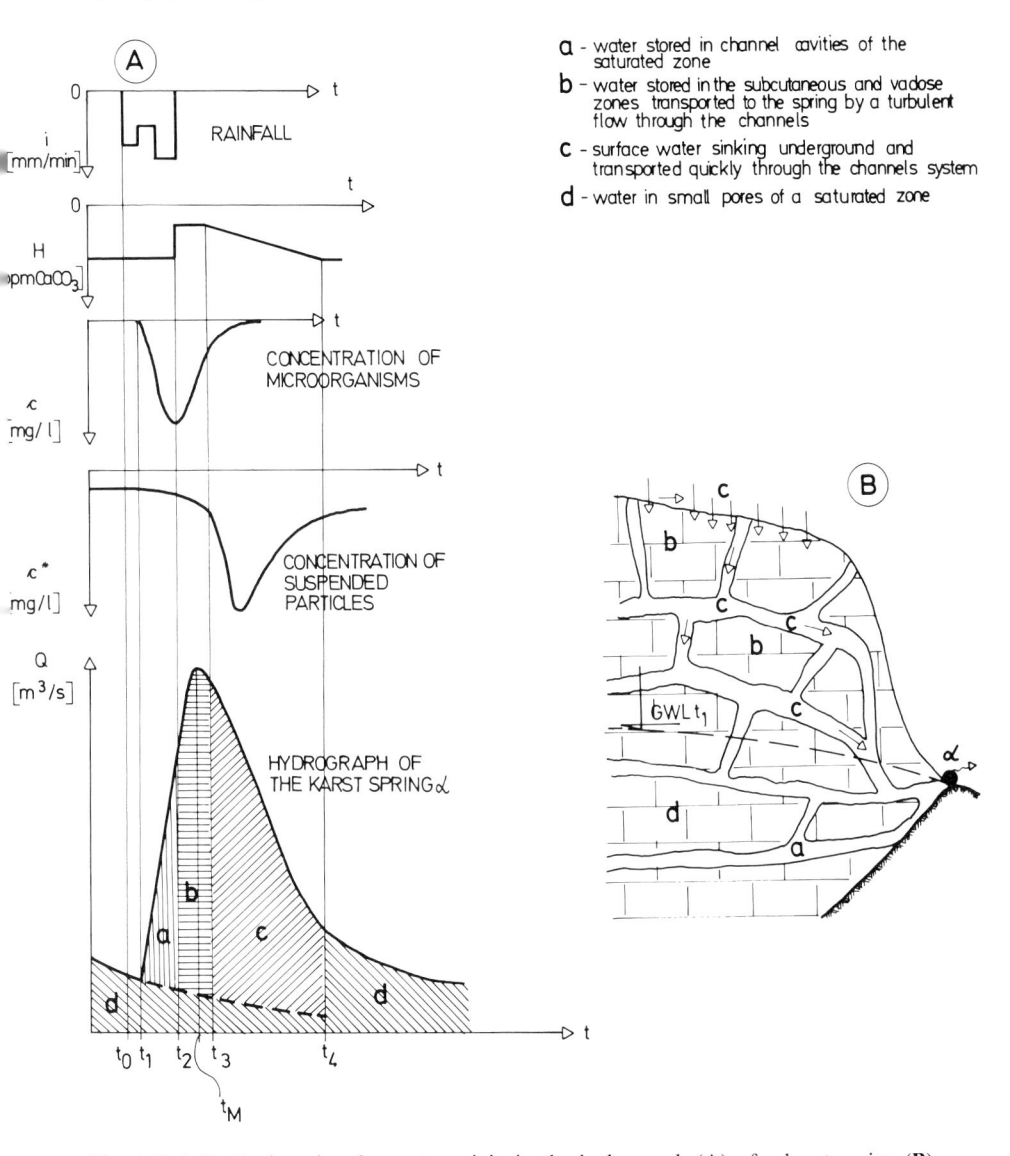

a - water stored in channel cavities of the saturated zone
b - water stored in the subcutaneous and vadose zones transported to the spring by a turbulent flow through the channels
c - surface water sinking underground and transported quickly through the channels system
d - water in small pores of a saturated zone

Fig. 4.27 A, B. Explanation for water origin in the hydrograph (**A**) of a karst spring (**B**)

ty. Atkinson (1977) and Williams (1983) similarly explained the hydrograph of a karst spring with regard to the areas that bring about its formation. The fissures of the karst massive resulting in effective porosity can be divided into four zones, with respect to their size (i.e. the velocity of the groundwater circulation in them) and whether they are located in a vadose or phreatic zone: (1) the zone of channel shafts or fissures of greater size, permanently located under the groundwater level, i.e. in the phreatic part of the karst massive (Zone *a* in Fig. 4.27); (2) smaller fissures, but closely connected, located in the vadose and subcutaneous zone

(Zone b, Fig. 4.27), where the transport is effected by turbulent flow; (3) the zone of channels and larger, well-connected fissures, located in a vadose zone; the water is retained there for a very short time due to a great flow velocity (Zone c, Fig. 4.27); (4) the zone of small fissures belonging to a phreatic zone where the flow is mainly laminar (Zone d, Fig. 4.27). Although this classification has certain drawbacks, it makes it possible to identify various types of water transport and to determine the fissure volume (at least approximately) in the area surrounding the karst spring. Figure 4.27 A is a graphic presentation of an explanation of the water origin in a hydrograph of a karst spring from the four mentioned zones. The hydrograph resulted from the rainfall starting at time instant t_0. The reaction of the system, i.e. the hydrograph increase starts a little later at time t_1. In the period of time between t_1 and t_2 the hydrograph is formed by water from Zone a, i.e. in the channels of the phreatic zone (area a). According to Atkinson (1977), the end of emptying for zone a can be exactly defined by measuring the total hardness, and according to Bögli (1980) by measuring the temperature of the spring water. The hardness of the water stored in the underground for a longer time is greater than that of the water coming from the atmosphere and sinking to the groundwater level by fast percolation. The temperature of the water retained underground for a longer period of time is constant and essentially differs from the temperature of water reaching the phreatic zone by fast percolation. The water volume obtained by integrating the hydrographs (subtracting the volume of the extrapolated part of the recession curve) from t_1 to t_2 corresponds to the volume of the shaft system in the phreatic zone. When the water hardness at the spring rapidly decreases or when the water temperature suddenly changes (more often increasing than decreasing), i.e. at time t_2, it represents the inflow of the new quantities of water stored in the subcutaneous and vadose zones. The hardness of that water is due less to its mixing with the water coming from the atmosphere. The longer the water is retained, the greater its hardness until it reaches the initial values. A similar situation is valid for temperature. The end of emptying of water stored in the vadose and subcutaneous zone begins approximately at time t_3. It is then that the number of the suspended particles in water, i.e. its muddiness, suddenly begins to increase. This number actually starts to increase a little earlier at time t_1, but the trend of rising is slow particularly from t_1 to t_2 and then from t_2 to t_3. When studying and analyzing the spring hydrograph, significant influence can be effected by measuring the concentration of the microorganisms in the spring water. The karst underground is abundant in microorganisms of animal origin, developing in water, prevalently in the flooded karst areas, i.e. in the phreatic zone. The curve of the change in the concentration of microorganisms in water shows their disappearance and increase from time t_1 to t_2, the beginning of the concentration decrease in time t_2 and almost complete disappearance in time t_3. If two rainfalls follow one another, microorganisms are not to be expected in the second water wave. The reason is simple and can be explained by the fact that the channel system was washed free of microorganisms during the first water wave, and there was no time for their return and recovery until the occurrence of the second wave. Starting from time t_3 the spring hydrograph is formed mainly from the water sinking underground from the surface, and it is transported by a rapid flow through the system of privileged paths, i.e. by underground channels and shafts. The end of the groundwater inflow from Zone *c* is expressed by the time instant t_4. From that moment onward, until the occurrence of a new rainfall, the spring is formed by the water stored in the phreatic zone, especially in its small fissures.

It should not be particularly stressed that this explanation has been simplified, i.e. it refers to a simple, almost ideal, and, consequently, small karst system. More or less significant changes occur when the water flows to the spring from several horizons, i.e. poljes in the karst, or when one part of the catchment is bare and the other covered by vegetation. The explanation presented in Figure 4.27 is not valid either when there are several systems of inflow and water storage within a large karst system. However, even in these more complex situations, the system can be identified similarly. In that case, it is necessary to carry out numerous, well-organized measurements of various quantities, particularly the chemical content of water, its temperature and the hydrologic-hydrogeologic-hydraulic parameters. Specific flood hydrographs have to be used in order to perform the above mentioned analyses. The most appropriate hydrographs would be those which are formed by intensive heavy rainfalls after long dry periods. In that case, all the described phenomena will be clearly expressed. Consequently, a great number of such hydrographs selected over several years of measurements should be studied and then the data should be statistically analyzed and only then physically interpreted.

4.4 Determination of the Catchment Area

The determination of the catchment boundaries and the catchment area is the starting point in all hydrologic analyses and one of the essential data which serve as a basis for all hydrologic calculations. The definition of the catchment areas or groundwater recharge areas for karst springs is important for estimating the groundwater supplies and for identifying the possible sources, directions and velocities of the contaminant movement. For water circulation in karst, this classical hydrologic problem represents a very complex, sometimes not easily solved, task. Only a few terrains in karst have been studied well enough as to make it possible to define the catchment precisely. In addition to obtaining topographic data on the terrain, it is necessary to carry out measurements which are used in the determination of the groundwater circulation under the different conditions of the groundwater levels. In order to exactly define the surface and subsurface catchment boundaries, it is necessary to conduct detailed geologic investigations and, accordingly, extensive hydrogeologic measurements. These measurements primarily include the existence of the connections (links) between individual points in the catchment area (connections: ponors-springs, piezometers-piezometers, piezometers-springs) applying one of the tracing methods: dye tests, chemical tests, solid floating particles or radioactive matter. The catchment areas in karst vary according to the groundwater levels, i.e. change with time. Only in exceptional cases do the surface and subsurface watershed lines coincide and only in those places where the boundaries between catchments are located in impermeable rocks. If this boundary is located in permeable carbonate layers it is

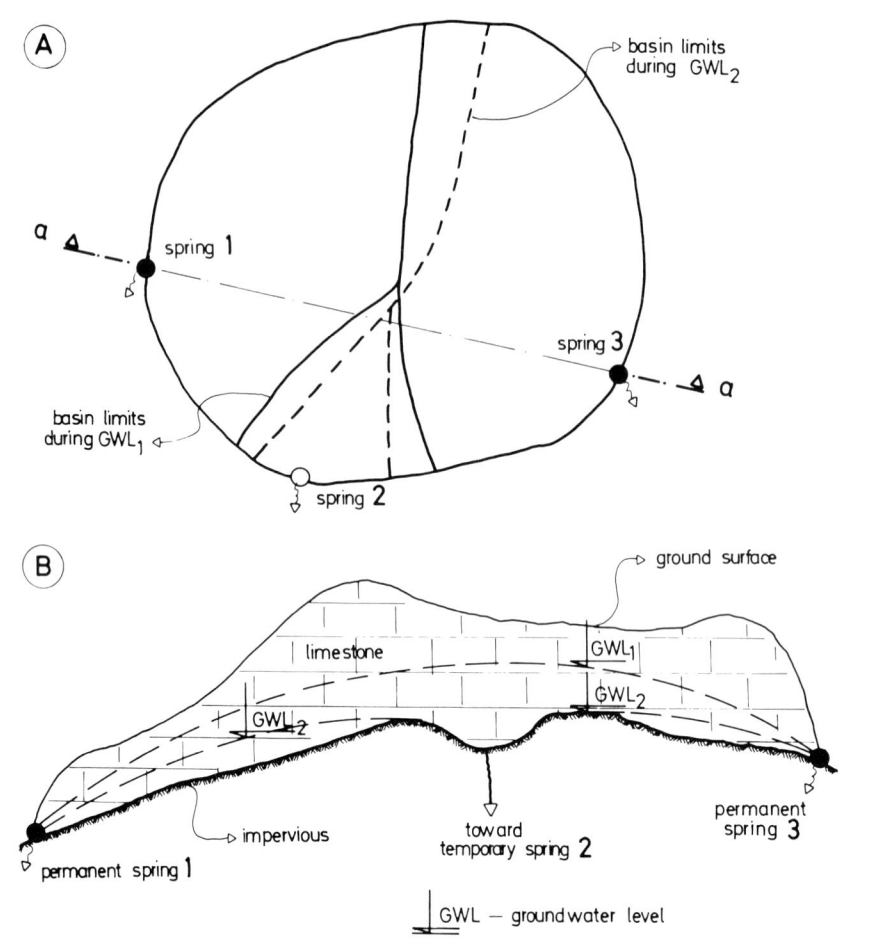

Fig. 4.28 A, B. Explanation of catchment areas of springs in the karst; situation (**A**); cross-section (**B**)

not stable. There is a certain zone within which the watershed limit is moved towards one or the other spring or towards the streamflow. The position of the watershed line depends upon the groundwater levels which change in time. In some situations at very high groundwater levels (after heavy rainfall) channels are activated in the karst underground thus causing the redistribution of the catchment areas. Figure 4.28 shows a schematic representation of this case. Figure 4.28 A shows the situation of three springs in karst with the catchment boundaries dependent upon the groundwater levels. When the groundwater level is high (case GWL_1) Spring 2 is active, and its catchment area significantly increases. At a low groundwater level (case GWL_2), occurring after long dry periods, the groundwater watershed line of Spring 2 moves inland so that its catchment area is considerably decreased. At the same time, Spring 2 stops functioning.

Figure 4.29 shows three cases outlining the relationship between a topographic (orographic) and hydrologic (hydrogeologic) spring catchments in the karst. In

Determination of the Catchment Area

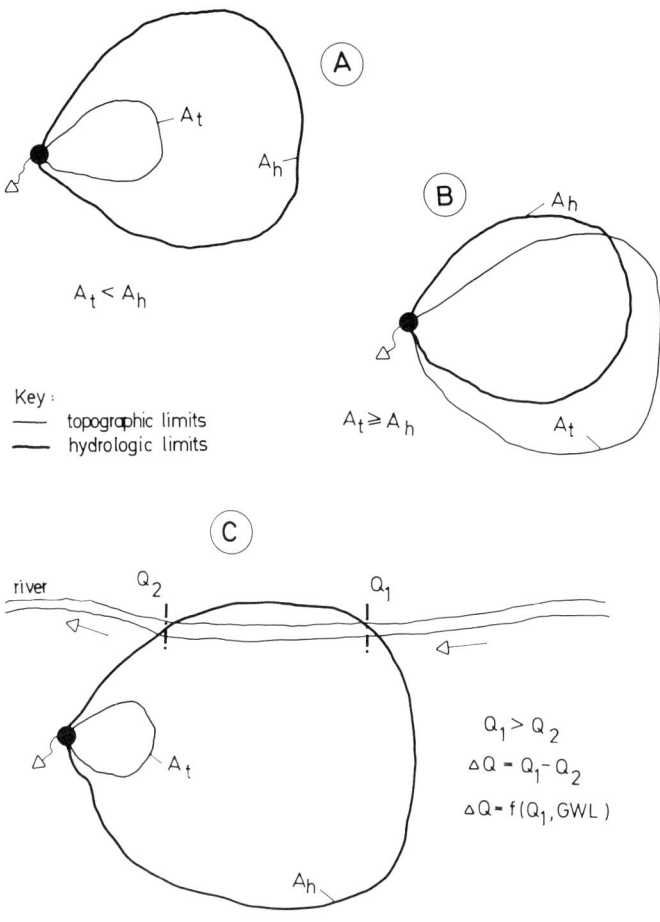

Fig. 4.29 A–C. Three relations between topographic A_t and hydrologic A_h catchment area for karst springs. (**A**) A_t; (**B**) A_h; (**C**) special case when a permanent streamflow is included in the spring catchment along one section

most cases the basic topographic catchment area A_t is smaller than the hydrologic area A_h, whose boundaries are located within the hydrologic catchment as shown in Figure 4.29 A. In practice, it is easy to determine the topographic catchment area, whereas the determination of the hydrologic catchment area is a complex task, difficult to carry out precisely and reliably. Čorović et al. (1985) reported that the ratio between the topographic catchment areas of the springs and the estimated hydrologic catchment area in the Dinaric karst (Yugoslavia) ranges from 1:2.3 to 1:70. Consequently, there are very complex situations in the natural surroundings, e.g. as shown in Figure 4.29 C. In that case, a permanent open karst streamflow is included in the hydrologic spring catchment along one section, and after a certain period of time leaves it. At the same time, there are permanent water losses into the karst underground along the given section caused by the cracks located at the bottom and banks of the streamflow. The water losses ΔQ

of the open streamflow along the section, flowing through the hydrologic catchment, depend upon the inflow discharge Q_1 and the groundwater levels (GWL). The water quantities ΔQ, i.e. the losses of the streamflow, feed the spring, and, thus, the spring catchment area is increased, whereas the open streamflow catchment area is decreased respectively. Consequently, it is evident that the determination of the spring catchment area presents a difficult task, since it changes in dependence on the groundwater levels as well as according to the discharge of the open streamflow intersecting the hydrologic spring catchment. This actual case refers to the Ombla Spring and the Trebišnjica River in the Dinaric karst of Yugoslavia. Naturally, the three previously described examples represent only a few of the most characteristic examples of the relationship between the topographic and hydrologic catchment area in karst. There are many more complex cases in practice. In the initial phase of such considerations, it should be stated that there are simple situations when only one part of the spring catchment area is covered by karst, and the topographic and hydrologic areas entirely overland. There are more common hydrologic cases identical with those in other more uniform aquifer which do not call for special analyses.

Dreiss (1983) stated that "linear kernel functions derived from the springflow response of large karst springs to intense, isolated storms are most physically realistic and have the greatest predictive accuracy when the assumed spring recharge area is consistent with tracer study results. This observation implies that, if sufficient precipitation and spring discharge data are available, such derived kernel functions could be useful for identifying or validating assumed spring recharge areas. In the absence of detailed tracing data a researcher might propose a number of feasible recharge areas for a spring and derive kernel functions, using the average precipitation for each of these proposed areas. The shape and predictive accuracy of the derived kernels would reflect the accuracy of the location of the assumed recharge areas. The effectiveness of these kernel function properties in delineating recharge boundaries would depend on the size of the spring relative to the areal extent of the monitored storm, the homogeneity of soil and vegetation conditions, and the area distribution of the precipitation stations". In his investigation Dreiss (1983) essentially applied the principles and theory of the direct outflow hydrograph under the conditions of the water circulation in karst. According to Dreiss "the major difference between methodology of kernel identification for springflow and for surface flow is related to the estimation of the input series. In surface runoff studies, the input series is usually an average of measured precipitation over the drainage as the volume of water available for runoff after evapotranspiration and soil-moisture conditions are taken into account. In springflow studies the input values are the moisture volumes per time increment which enter the aquifer during or after a storm and which contribute to the rapid storm response of a spring discharge. Mass conservation requires that the sum of these volumes equals the total volume of the storm response of the spring discharge".

Estimation of the groundwater recharge (Dreiss 1983) requires: (1) that the rainfall be averaged over the spring recharge area; (2) that excess precipitation be computed with a moisture balance technique; and (3) that the volume of rapid recharge be separated from the volume of excess precipitation. Evidently, a clas-

Determination of the Catchment Area

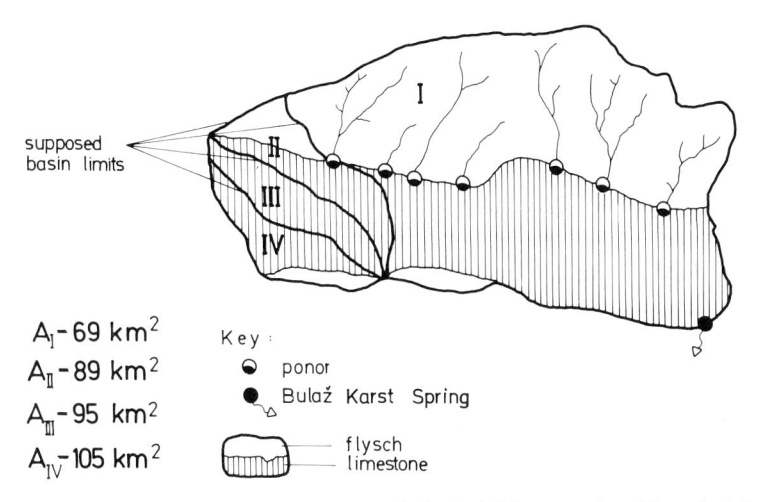

A_I - 69 km^2
A_{II} - 89 km^2
A_{III} - 95 km^2
A_{IV} - 105 km^2

Key:
- ponor
- Bulaž Karst Spring
- flysch
- limestone

Fig. 4.30. Supposed catchment areas of the Bulaž karst spring (Yugoslavia)

sical hydrologic method is applied so that the conditions of the groundwater flow to a spring in karst are adequately transformed and the input series of rainfall is estimated. Since these calculations involve the technique used for the direct runoff, hydrographs can be applied for the determination of karst spring catchments whose recharge is carried out by conduit flows. This method can be applied also for the outflow conditions occurring immediately after heavy rainfall when the diffuse flow is insignificant, and when the conduit flow by its characteristics (primarily velocity) resembles overland flow. Hence, it can be concluded that the described method is not recommendable for diffuse karst springs.

The example of the karst spring catchment Bulaž in Istra (Yugoslavia) will be used to illustrate the application of the hydrologic budget method for determining the catchment area of the karst springs. Figure 4.30 shows four possible alternatives of the catchment area determined by hydrogeologic investigations (Magdalenić et al. 1986). It is evident that one part of the catchment is covered by karst, and the other by flysch layers. A ponor zone into which the surface streamflows sink occurs at the places of contact between flysch and karst. The water from these streamflows emerges at the Bulaž Spring. The connection between the ponors situated close by and the springs has been proved by dye tests, whereas the dye injected into the farthest ponors disappeared. Hydrogeologists have stated four possible alternatives of the spring catchment areas covering 69 to 105 km^2. The spring outflow was measured only in 1980 and 1981, so that in this period it was possible to define the effective precipitation Pe emerging at the spring as outflow.

The quantity of rainfall P_i in the catchments i ($i \in 1$ to 4) was calculated by applying the Thiessen polygons. The runoff coefficient a_i for each of the four alternatives has been defined by the expression:

$$a_i = Pe/P_i \qquad (4.4)$$

Table 4.1. Runoff coefficient a_i for four alternative catchment areas of the Bulaž Spring

Year	a_1	a_2	a_3	a_4
1980	0.79	0.61	0.57	0.52
1981	0.85	0.66	0.61	0.56

and the result is shown in Table 4.1. In the next phase of the analysis it is necessary to use one of the hydrologic expressions (or more complex methods) in order to define the annual runoff deficit. The computation was performed using two equations, i.e. the Turc method (Eq. 4.5) and the Coutagne method (Eq. 4.6). Both equations yielded identical results.

The annual runoff deficit D expressed in mm, defined by the Turc equation, is as follows:

$$\left. \begin{array}{l} D = P(0.9+P^2/L^2)^{-0.5}; \\ L = 300+25\,T+0.05\,T^3. \end{array} \right\} \qquad (4.5)$$

Coutagne equations are:
Valid for

$$\left. \begin{array}{ll} (\frac{1}{8}\lambda) \leqslant P < (\frac{1}{2}\lambda); & D = P - P^2\lambda, \ \lambda = 1/(0.8+0.14\,T); \\ P < (\frac{1}{8}\lambda); & D = P; \\ p \geqslant (\frac{1}{2}\lambda); & D = 0.20+0.035\,T. \end{array} \right\} \qquad (4.6)$$

D and the annual precipitation P are expressed in mm; the mean annual temperature T is expressed in °C.

In the next phase four alternatives of the runoff deficit D_i^* are defined using the equation:

$$D_i^* = P_i - Pe \qquad (4.7)$$

whereby the effective precipitation Pe is a value which is constant and does not depend upon the variations in the catchment area. The effective precipitation represents the actual measured outflow at the spring. The time unit used in the application of this method is 1 year, and all the values are preferably expressed in mm or m³.

Table 4.2 shows the results of the runoff deficit calculated using Eq. (4.7) and is compared with the results obtained by the Turc method (Eq. 4.5). The values presented in Table 4.2 show that the most acceptable value of the catchment area is A_4 covering 105 km². For that area the runoff coefficient a is very real and ranges from 0.52 to 0.55.

The previously described budget method is applicable for 1 year as a unit of time. Since the method is not very sensitive, its results are not quite reliable. It should be applied only when it has been proved that the Turc (Eq. 4.5) and Coutagne (Eq. 4.6) models yield real results of the runoff deficit. The hydrologic budget method should be used only for preliminary investigations into the magnitude of the spring catchment area in karst. This method should include the

Determination of the Catchment Area

Table 4.2. Difference in the runoff deficit for the four alternative catchment areas of the Bulaž Spring

Year	$\Delta_i = D_i^* - D_i$							
	$A_4 = 105$ km²		$A_3 = 95$ km²		$A_2 = 89$ km²		A_1 69 km²	
	[10⁶ m³]	[%]	[10⁶ m³]	[%]	10⁶ m³	[%]	[10⁶ m³]	[%]
1980	0.67	1.0	4.3	6.4	8.3	12	22	32
1981	9.10	14	14	22	17	27	28	43

regional expressions for the definition of the annual or seasonal outflow deficits where they already exist. The accuracy of the method is influenced by the selection, position, number and accuracy of the measurements carried out at the rainfall gauging stations located in the given catchment area. If the number of gauging stations is small and their distribution unfavourable, this method is not likely to yield reliable results.

A groundwater hydrograph describes the outflow from a groundwater reservoir into an open watercourse. This outflow may be due to recharge of the reservoir by rainfall, or may be caused by the release of groundwater from the storage in the absence of rain, or by a combination of both effects. Theoretical treatment of the groundwater hydrographs is based on the theory of the non-steady flow of the groundwater in which changes in the storage control the flow. The groundwater hydrograph is important for the definition of the relation between the rainfall and the spring discharge if there is no surface runoff.

For practical application, the following is a system of Eq. (4.8) presenting a simplified theoretical solution for the definition of the groundwater hydrograph:

$$\begin{aligned} Q_T &= f_1(1-e^{-T/j}); \\ Q_{2T} &= f_2(1-e^{-T/j}) + e^{-T/j} Q_T; \\ Q_{iT} &= f_i(1-e^{-T/j}) + e^{-T/j} Q_{(i-1)T}; \\ Q_{NT} &= f_N(1-e^{-T/j}) + e^{-T/j} Q_{(N-1)T}, \end{aligned} \qquad (4.8)$$

where Q_{iT} is the discharge hydrograph of the groundwater expressed in m³ s⁻¹ or mm for the chosen time unit in time instant iT; T is the time increment of the computation which can be 1 day or one part of the day, depending upon the catchment area and the available input data; j is the reservoir coefficient expressed as a time unit; whereas, f is the groundwater recharge consisting of the sum (difference) of precipitation P_i, evapotranspiration ET_i and the soil moisture deficiency SM_i according to the equation:

$$f_i = P_i - ET_i - SM_{i-1}. \qquad (4.9)$$

If f_i is smaller than zero, the computation used the value zero. The dimension taken for all the parameters given by Eq. (4.9) is either m³ s⁻¹ or mm/time.

According to Eqs. (4.8) and (4.9), it is evident that in order to define the groundwater hydrograph applying this method, the following hydrologic or meteorologic characteristics of the analyzed catchment should be considered: (1) precipitation P; (2) evapotranspiration ET and (3) soil moisture deficiency SM. The accuracy of the calculation is strongly influenced by the selection of the precise value for the reservoir coefficient j.

The procedure used for the definition of the spring catchment area in karst involves the calculation of the spring hydrograph by expressions (4.8) and (4.9) and by comparing the obtained hydrograph with the measured hydrograph by varying the spring catchment area. The exact catchment area of the spring in karst can be defined only after achieving an optimum agreement between the hydrograph obtained by measurements and the hydrograph calculated using the suggested model. Since this model has not been developed for the outflow in karst, but for a more homogeneous medium, its application to the karst terrain exhibits some specific characteristics. Hence, these characteristics should be constantly taken into account. In some cases, the model will not yield acceptable results. This primarily refers to the karst conduit springs, and so, the model should be used only for the diffuse karst springs or for the definition of the recession parts of hydrographs.

Since the recharge will occur only after the replenishment of the soil moisture deficiency, it is necessary to study the soil moisture content for each analyzed catchment. In karst situations the data on the groundwater levels should be used for the above purpose. These data are also significant for the estimation of the real evapotranspiration. Soil evaporation depends upon the supply of moisture to the unsaturated zone by infiltration from precipitation and by capillary rise from the groundwater table. If the unsaturated zone remains at field capacity, the evaporation from a bare soil may be the same as the potential evapotranspiration. If the groundwater table is close to the surface, the soil evaporation may be about equal to free water evaporation. If the water supply is only by capillary rise from the groundwater table, then the rated soil evaporation will be governed by the depth of the groundwater table and the soil properties. Evaporation decreases with increasing depth. Even in heavy soils evaporation will cease when the water table is at a depth greater than about 2 m below the surface. In karst, in general, and particularly in the bare karst, the variations in the groundwater levels are very rapid and occur with great oscillations. The groundwater level is often situated far below the soil surface. In such cases, the soil evaporation should be considered as zero. The transpiration depends upon the vegetation cover, so each catchment has to be studied in detail in order to determine it exactly. A formal approach to the determination of the soil moisture deficiency and evapotranspiration for the catchments in karst cannot be recommended since a systematic error can be introduced into the discussed model.

The parameters of the soil moisture deficiency SM and the reservoir coefficient j can be defined also using the optimization procedure built into the program developed for the electronic computer. It is sufficient to use daily data on the hydrologic and meteorologic characteristics when dealing with catchments covering an area greater than 100 km^2. Shorter time periods should be used when dealing with the catchments covering smaller areas. Since it takes a certain period of time for the rainfall water to reach the spring, the recharge has a time lag with respect to the rainfall. This time lag and, consequently, the reservoir coef-

Determination of the Catchment Area

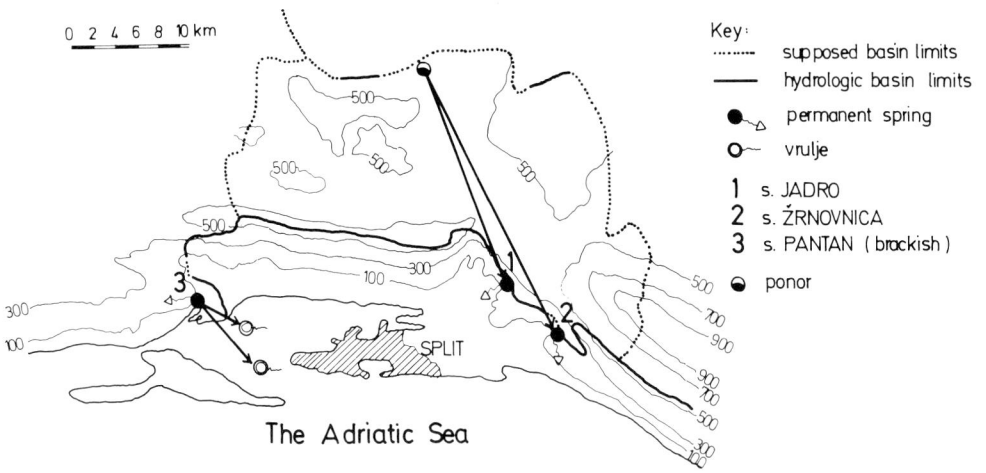

Fig. 4.31. Map of three karst spring catchments in the vicinity of Split (Yugoslavia)

ficient differ, particularly after heavy rainfall in the wet period, from those occurring after rain of low intensity in the period of low groundwater levels. In the first case, a turbulent flow through privileged paths is dominant, whereas in the second case, the water reaches the spring by a slow diffuse flow. In order to determine the spring catchment area, it is necessary to use the suggested model to simulate the hydrography over at least one entire year.

The groundwater hydrograph method has been used for the determination of the catchment areas for two springs in the bare Dinaric karst (Yugoslavia) in the vicinity of the city of Split, situated on the coast of the Adriatic Sea (Fig. 4.31). In spite of the numerous and detailed hydrogeologic investigations carried out, it has not been possible to define even approximately either the catchment boundaries for the two springs or the exact catchment area. The spring of the Jadro River is very important since it has supplied the Split with water since Roman times (150 B.C.). It is located at an elevation of 30 m above sea level. The water from the Žrnovnica Spring emerges at the surface at different elevations, from 78 to 90 m above sea level, depending upon the groundwater levels in the hinterland of the karst mass. The higher the groundwater levels, the higher is the elevation where the Žrnovnica Spring emerges. The mean annual discharge of the Jadro Spring for 1981 was $8.51 \text{ m}^3 \text{ s}^{-1}$, and of the Žrnovnica Spring $2.07 \text{ m}^3 \text{ s}^{-1}$. The determination of the catchment area has been carried out according to the data obtained during that year; one calculation was performed for the daily discharge of the entire year, and the second for the data referring to only 11 months. The month of December was left out of the calculations as the rainfall was particularly high that year, which resulted in high average monthly discharges of $25.4 \text{ m}^3 \text{ s}^{-1}$ (Jadro) and $5.90 \text{ m}^3 \text{ s}^{-1}$ (Žrnovnica). The conduit flow was dominant in December and due to the increase in the groundwater levels great quantities of water were stored underground, which evidently influenced, to a certain extent, the definition of the catchment area. Table 4.3 shows the results of the calculation related to each catchment and to both catchments of the karst springs: Žrnovnica

Table 4.3. The catchment area A, the reservoir coefficient j and the standard error S for the Jadro and Žrnovnica Springs

Spring	I–XII 1981			I–XI 1981		
	A [km^2]	j [day]	S [m^3 s^{-1}]	A [km^2]	j [day]	S [m^3 s^{-1}]
Jadro	257	10	5.31	227	16	3.15
Žrnovnica	48	8	1.23	53	9	1.16
Jadro + Žrnovnica	297	8	5.12	286	13	3.60

and Jadro. In addition, it states the optimal reservoir coefficient j, and the standard computation error S, defined by the expression:

$$S = \sqrt{\frac{\sum_{i}^{N}(Q_{m_i}-Q_{c_i})^2}{N}}, \qquad (4.10)$$

where Q_{m_i} is the discharge obtained by measurements carried out in day i; Q_{c_i} is the discharge calculated using the model; and N is the number of days for which the computation was carried out.

According to the data from Table 4.3, it can be stated that the standard errors are relatively high and that they are ca. 50% of the average annual spring discharge. Their value is significantly influenced by high discharge, whereas the agreement between the measured and calculated discharge is exceptional in the recession period. If the model is applied only for the very wet December 1981, the hydrologic catchment area of the Jadro Spring will be 260 km^2, and of the Žrnovnica Spring 40 km^2. The results of the computation illustrate that the variations in the catchment area are dependent upon moisture. It should be noted that the reservoir coefficient j is significantly greater in the dry period. For December only, its value is $j = 3$, thus making it evident that the conduit flow is dominant. In order to carry out the computation as accurately as possible, only the hydrologic and not actual years should be taken into account. The determination of the precise locations of the catchment boundaries is the subject of further hydrogeologic investigations, since that task could not be carried out even approximately by including the data obtained by gauging stations into the computation. The number of gauging stations was insufficient, and the precipitations were relatively uniformly distributed along the analyzed area.

The determination of the catchment area in karst should not be exclusively limited to the springs. The same or similar methods can be used to define the catchment areas up to any gauging station in the open karst watercourse. Subsequently, the application of the Stanford Watershed Model (SWM) to the catchment of the large karst in the Cetina River in the Dinaric karst (Yugoslavia) will be described. SWM is primarily a computer model intended for the simulation of a great part of the surface and subsurface phases of the hydrologic cycle in the entire catchment. SWM has been classified as a simulation model which continuously changes in the state and has a great number of parameters. The input data include the data on rainfall, evapotranspiration, a large number of hydraulic parameters and catchment parameters. The modelling of the hydrologic cycle starts with precipitation which had been retained and infiltrated into the catchment. One part of the retained water is distributed into the area of the upper and lower soil zones. The upper zone controls the surface flow and the subsurface outflow. The lower zone regulates the long-lasting infiltration, i.e. the reserves of the groundwater, and this forms the base of the hydrograph. One part of the water quantity from the upper zone, lower zone and underground water storage is also lost by evapotranspiration. Finally, the combination of the surface, subsurface and underground flows results in a total simulated flow. This sequence of computations can be applied to numerous catchments and subcatchments. In the modelling process, the calibration and the control testing of the model and validity are based upon the existing data on rain and water discharge. The original SWM has been extended into numerous versions in order to adapt it to specific tasks. The version used in this book was developed at the National Weather Service. The computer program of the NWS version of the SWM was written in FORTRAN.

SWM was tested (Roglić et al. 1985) on the Cetina Catchment up to the Vinalić gauging station, with the respective topographic catchment area covering 199 km^2. The selected year was 1982/1983 when several large water waves were observed, and for which climatologic and hydrologic data of a satisfactory quality and quantity were available. The basic hydrologic data used to simulate the outflow in a catchment include: potential evapotranspiration, precipitation and average daily discharge at the outflow profile during the chosen period. In order to apply the model consistently, it was necessary to include the data on rainfall obtained at ombrograph stations, since the input data were rainfalls lasting 6 h.

In karst terrains, as the Cetina Catchment, the outflow deficits are small, primarily due to the fast sinking of water underground. Consequently, special attention should be paid to the determination of losses by evapotranspiration. In the preliminary phase it was necessary to define the concentration time, as well as the relation time-area. The program IZOKRON (Bonacci and Roglić 1982) was used for the definition of isochrones and for the computation of the concentration time and areas between isochrones. The time obtained for the catchment concentration, related to the Vinalić Gauging Station, is 26 h. The first alternative, tested for the topographic catchment area, showed that it is not possible to achieve a satisfactory outflow simulation without regard to the choice of input parameters. The simulated discharges were constantly significantly smaller than the measured discharges. Satisfactory results were obtained only after studying the catchment area of 470 km^2, wherefrom it can be concluded that the hydrologic catchment area is even 2.4 time greater than the topographic catchment area itself. Thus, the size of the catchment area was obtained, but not its location in space. In order to determine the actual watershed line it would be necessary to know the geologic and hydrogeologic factors influencing the direction of the groundwater circulation. According to the previously described investigations, the position of the watershed line is not likely to be constant; on the contrary, it depends upon the water levels in the karst underground. After heavy rainfall in a flooding period, springs appear along the Cetina which drain the water from the higher karst poljes. In that period the catchment boundary is located farther away from the

Cetina. During the periods of low waters, the water is lost to the Jadro Spring (as proved by dye tests) and the catchment boundary approaches the Cetina River. As the hydrologic year 1982/83 can be considered a wet (or even exceptionally wet) year, the estimated catchment area of 470 km^2 is supposed to be close to the maximum hydrologic area. The testing of the model sensitivity to numerous parameters, i.e. the influence of the input parameters on the changes in the outflow volume or the hydrograph peaks, has shown that there are three basic groups of parameters. The first group includes the parameters whose changes slightly influenced the results, e.g. the evapotranspiration parameter for the humidity of the upper zone $K3$, the portion of inflow into the underground caused by deep seepage $KK24$ and a part of the impermeable area $IMPV$. The second group includes the greatest number of parameters, although their influence on the results could not be clearly defined due to their interdependence. The last group includes the most important parameters, whose changes, even the slightest, influenced the results, and whose connections were clearly defined: the Manning roughness coefficient NN, the infiltration index CB, the average length of flow along the terrain L and the index of water quantity entering the subsurface flow with regard to the surface flow CB. In this actual case, the greatest deviations appeared between the peaks of the simulated and measured hydrographs, with the peaks of the simulated hydrographs being continually smaller than the peaks of those measured. Therefore, the parameters CB, CC were minimized to the lowest value with regard to the first alternatives. The NWS version of the Stanford Model does not include the influence of snow on the outflow. Consequently, in the winter and spring months, which showed real, significant deviations between the simulated and measured discharges (due to precipitation, i.e. snow), a physically justified redistribution of precipitation was performed, depending upon the air temperature.

The testing of the model in the catchment up to the Šilovka Gauging Station showed that there is a great disproportion between the topographic and hydrologic catchment areas as well as near the Vinalić Station. The topographic area of the catchment was 594 km^2, and the area which, according to the Stanford Model, approximated best the natural flow, covers 1440 km^2. The Šilovka Station was used, in a way, to control the parameters chosen for the hydrograph simulation at the Vinalić Gauging Station. In other words, the combination of parameters taken as optimal for the Vinalić Station was used also for the flow simulation of the Šilovka Station with some slight corrections. Since the catchment area is considerably greater, the analysis included new rainfall gauging stations. The redistribution of the snow precipitation was carried out in the spring months, depending upon the air temperatures. In addition, the parameter CB (index of the infiltration quantity) was increased from 0.31 (chosen for the Vinalić Station) to 0.50, since it was shown that the peaks of the simulated hydrograph for the combination of parameters, chosen for the Vinalić Station, were somewhat higher than the measured ones. Taking into account this correction, the chosen combination of parameters approximated the flow in the riverbed quite well. The values of the measured and simulated hydrographs were plotted in Figure 4.32 A and B. Figure 4.33 illustrates the situation in the Cetina Catchment with the plotted topographic and supposed hydrologic catchment boundaries. The contact of

Determination of the Catchment Area

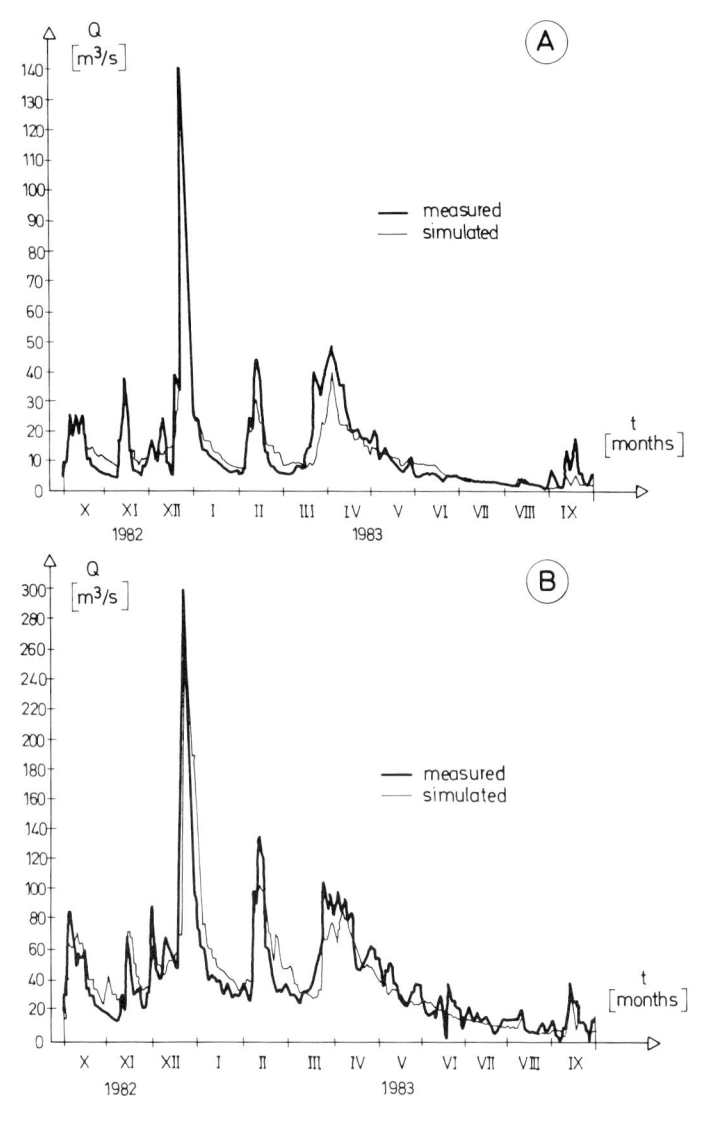

Fig. 4.32 A, B. Simulated and measured hydrographs of the Cetina River at the Vinalić (**A**) and Šilovka (**B**) gauging stations (Yugoslavia)

the upper horizons, lying outside the topographic catchment, with the open flow of the Cetina, is effected via numerous permanent and temporary springs located prevalently on the left bank of the Cetina River. All those springs are situated either in the close vicinity of or in the open streamflow of the Cetina itself.

The links between the open streamflow on the lower and higher horizons in karst are a common phenomenon. They cause numerous problems from the hydrologic and hydrogeologic standpoints and call for extensive and expensive theoretical and practical investigations. The case of the Zrmanja River is par-

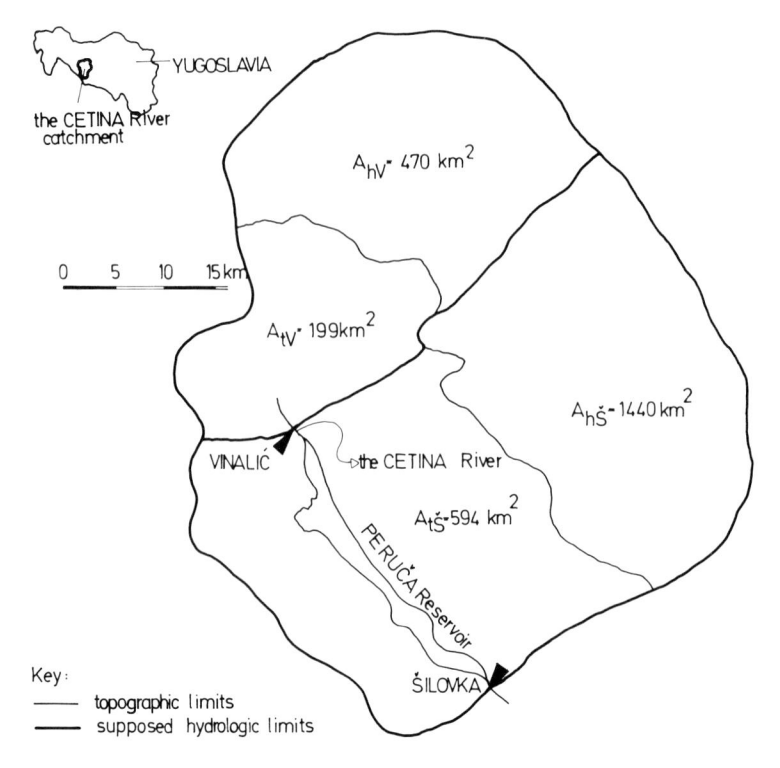

Fig. 4.33. Map of topographic and supposed hydrologic limits on the Cetina River Catchment (Yugoslavia)

ticularly interesting from the standpoint of karst hydrology. Numerous springs and ponors in the close vicinity of or in the very streamflow influence significant variation of the catchment area along its course. The relationship between the water of the upper horizon I (Fig. 4.34) and the water of the streamflow, as well as with the direct topographic catchment of the Zrmanja has been relatively well studied, but some facts have not as yet been explained. Numerous dyeing tests were carried out and many qualitative connections were established (Fig. 4.34). The primary objectives were: (1) to analyze the water quantity along the Zrmanja River; (2) to establish the river sections with underground losses and (3) to define the quantity of losses and inflows on particular sections of the streamflow. Accordingly, five water gauging stations were chosen and their essential data are presented in Table 4.4. According to the data obtained by the five rain gauging stations, the Thiessen polygons were defined and yearly precipitation P on the catchment was calculated. The analysis covered the period from 1953–1980. The next stage included the estimation of the yearly discharge deficit in particular topographic catchment areas, as covered by the chosen gauging stations. The discharge deficit is defined according to expressions (4.5) and (4.6). Effective yearly precipitation Pe can be determined by:

$$Pe = P - D. \qquad (4.11)$$

Determination of the Catchment Area

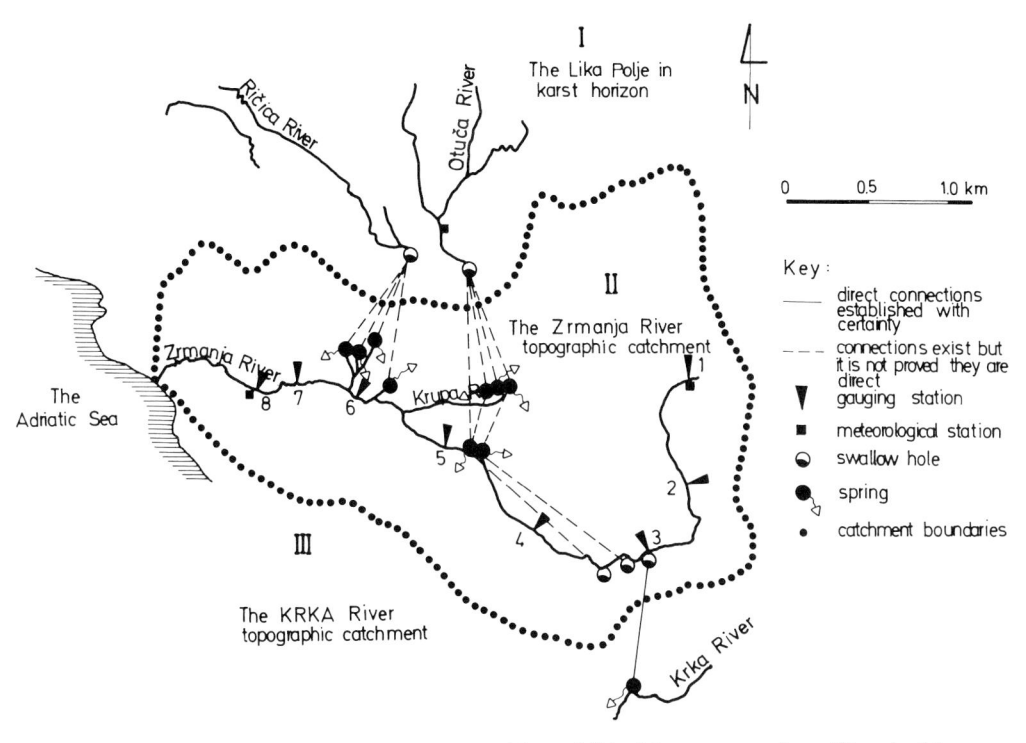

Fig. 4.34. The Zrmanja River Catchment with established karst connections (Yugoslavia)

Table 4.4. Essential data of five gauging stations in the Zrmanja River Basin

Gauging station	Established	Actual level [m a.s.l.]	Topographic catchment areas [km^2]
2	29. VIII 1952	254.903	156
3	8. V 1952	195.611	227.5
4	1. IX 1904	123.240	295.9
5	4. IV 1905	49.328	382.5
7	6. V 1952	3.064	665.5

This precipitation should flow out if the catchment area A has been exactly defined. As the catchment area is in karst, the area A given in Table 4.4 is not hydrologic, but just topographic. The comparison of Pe obtained by this analysis with the actual discharge quantities on the five analyzed gauging stations reveals the type of relationship between the topographic and hydrologic catchment areas. These data serve as an indicator of the influence of karst on particular sections of the catchment. The influence of karst on the "irregularity" of discharge has been further investigated by establishing the difference between the discharge as defined by field measurements and the discharge defined by Eq. (4.11). The discharge can be expressed either as actual precipitation in mm or in m^3 s^{-1}. Table 4.5 shows the final result of the analysis. The value ΔQ represents the difference

Table 4.5. Relation between measured mean discharge $\bar{Q}$ and calculated mean discharge $\bar{Q}_T$

Gauging station	$\bar{Q}$ [m³ s⁻¹]	$\bar{Q}_T$ [m³ s⁻¹]	$\Delta Q = \bar{Q} - \bar{Q}_T$ [m³ s⁻¹]	$\Delta Q/\bar{Q}$ [%]
2	6.06	5.03	1.02	+17
3	5.32	7.16	−1.84	−36
4	5.18	8.13	−2.95	−57
5	10.28	10.44	−0.17	−2
7	39.91	19.00	20.91	+52

between the discharges measured at the gauging station and those determined by Eq. (4.11), i.e. those related to the topographic catchment area. These results present the average discharges over several years (1953–1980) as defined by measurements $\bar{Q}$ and the Turc expressions (4.5) $\bar{Q}_T$. It should be stated that the computations were carried out taking the year as a time unit which directly points to the order of magnitude of the accuracy and sensitivity of the whole analysis. The average discharge defined by measurements on profile 2 is greater than the discharge obtained for the catchments in the non-karst terrains. Consequently, it can be concluded that the actual area covered by the hydrologic catchment of Station 2 is greater than its topographic area fo ca. 15%, i.e. approximately 180 km². This conclusion can be confirmed by the following facts: there are no significant losses on the section from Zrmanja Springs to Station 2, the piezometric level of the groundwater is above the level of the open streamflow during the whole year and the inflow into the Zrmanja Springs comes not only from the topographically defined catchment, but from the adjacent higher horizons. A negative difference amounting to −1.84 m³ s⁻¹ has been established at profile 3. It can be explained either by significant losses caused by water sinking through the "sieve" into the wetted area or through the swallow hole, or by the fact that the topographically defined catchment is greater than the actual hydrologic catchment. The first assumption, however, seems more probable since the piezometric level is lowered below the riverbed of the Zrmanja open streamflow just somewhere along the section from 2 to 3. Thus, the flow becomes "suspended" thereby influencing the water losses along this section. The difference established on profile 4 was even −2.95 m³ s⁻¹, which is an average of ca. 60% of that profile discharge. The situation found on the section from 2 to 3 is even more evident on the section from 3 to 4. There the flow is completely suspended, i.e. the groundwater level is essentially lower than the level of the riverbed of the Zrmanja open streamflow. It has not been discovered where these sinking waters appear. According to some assumptions most of them return to the Zrmanja riverbed just downstream from profile 4, and along the entire section downstream from profile 4. According to the second hypothesis, the greatest quantity flows to the Krka Catchment (Horizon III on Fig. 4.34). The most recent dye tests in 1985 confirmed this second hypothesis (Fig. 4.34). Presently, it has not been possible to define the exact quantitative relations. The difference on profile 5 was −0.17 m³ s⁻¹. This difference is statistically insignificant, and it can be stated that there exists a balance between the hydrologic and topologic catchment on that section. On profile 7 the established difference was ca. 21 m³ s⁻¹, i.e. about 50% of the measured mean discharge over several years. This quantity definitely proves that profile 7 covers an area greater than its topographic area, i.e. 300–350 km². This phenomenon can be physically explained by the connection established between the high horizon I and the Krupa Springs and some Zrmanja Springs.

4.5 The Origin of Brackish Karst Springs

This section discusses the mechanism of water salinization of the karst springs. From the standpoint of water supply, this problem is quite significant as it is still not possible either to explain it theoretically or to solve it practically. These great quantities of water cannot be used either as drinking water or for industrial and agricultural purposes. According to Yugoslav standards, the salinity of the drinking water is maximally 250 mg l⁻¹ Cl⁻. Some Mediterranean countries, as well as numerous countries with dry and warm climates, have raised this limit to 500 mg l⁻¹ Cl⁻. The inhabitants of the Sahara can drink water with a salinity of up to 2000 mg l⁻¹ Cl⁻. The problem of the exploitation of the salt water for irrigation depends upon the type of plant and the geomechanical and pedologic characteristics of the soil. Certain plants can withstand a fairly high percent of salinity, whereas other plants require great quantities of fresh water. The soil irrigated by brackish water should be loose and very permeable so that the salt is entirely washed off by the fresh water precipitation in the area.

Engineers and scientists have been dealing with this problem for a long time, since water is of special importance in the coastal regions and, primarily, in those areas where there is a water shortage. Among numerous papers on the relationship between brackish and fresh water in the coastal regions, the most significant, from the theoretical and practical standpoint, are the conclusions reached by Ghyben (1889) and Herzberg (1901). Their conclusions are based upon the laws of hydrostatics and on the relationship between the density of fresh ϱ_f and brackish waters ϱ_s. The density of the brackish water ϱ_s in kg m⁻³ is greater than the density of fresh water floating as a lense above the brackish water. Under hydrostatic conditions, the following expression is valid:

$$(h_1/h_2) = [\varrho_s/(\varrho_s - \varrho_f)], \qquad (4.12)$$

where h_1 is the depth of fresh water below sea level, h_2 is the height of fresh water above sea level. The explanations for Eq. (4.12) are shown in Figure 4.35. According to this law, the elevation of fresh water below sea level h_1 depends upon the elevation of the fresh water above sea level. With the increase of h_2, h_1 is increased, and with its decrease the depth of h_1 decreases. Since the difference $\Delta\varrho = (\varrho_s - \varrho_f)$ is constant for each location, the ratio h_1/h_2 ranges from 28 to 42 and most frequently from 36 to 40. If the elevation of fresh water above sea level is lowered by pumping, the depth of the lense of fresh water below the sea will be decreased accordingly, and the brackish water can penetrate into the kernel of fresh water. This phenomenon can occur on smaller islands or along the seacoast, and it can lead to the unpleasant consequence of the salinization of the lense of fresh water which can be removed only after a long period of time.

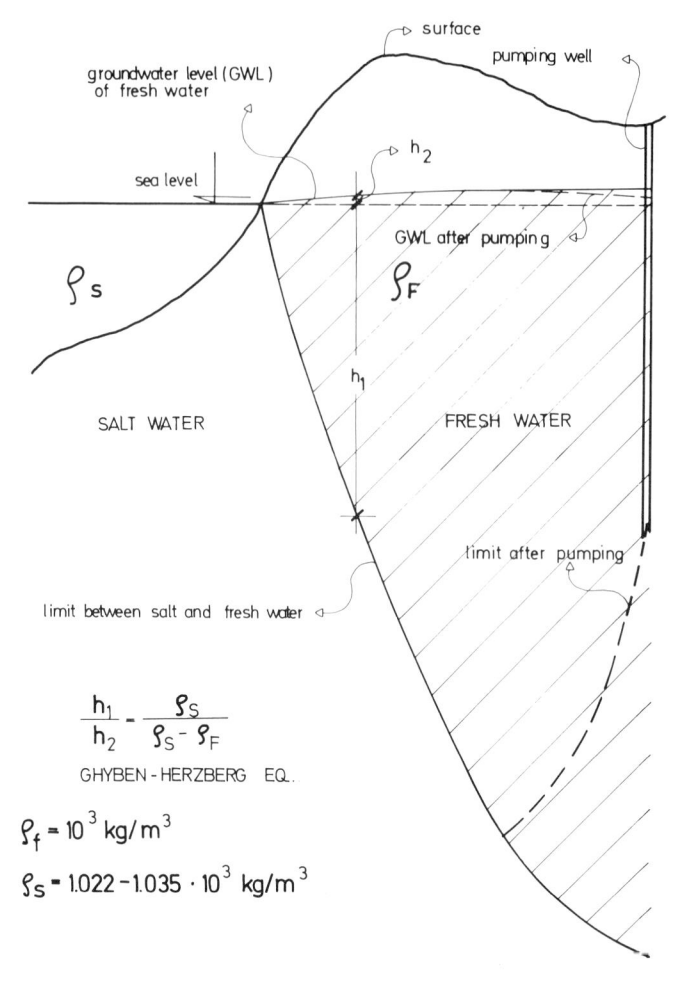

Fig. 4.35. Relationship of salt and freshwater according to hydrostatic law

$$\frac{h_1}{h_2} = \frac{\rho_S}{\rho_S - \rho_F}$$

GHYBEN-HERZBERG EQ.

$\rho_f = 10^3 \text{ kg/m}^3$

$\rho_s = 1.022 - 1.035 \cdot 10^3 \text{ kg/m}^3$

The relationship between fresh and brackish water in carbonate rocks in coastal areas was studied by Stringfield and LeGrand (1969 and 1971). They stated that: "If the sea level remained at a constant level and if the recharge from rain were at a uniform rate, the interface of the salt water with the fresh water would remain in the same position and the salt water would remain almost motionless while the fresh water would move toward the sea in the same way as if the hinterland were an impermeable surface". The Ghyben-Herzberg Law was formulated exclusively on the basis of hydrostatic considerations, and its reliability is limited. The dynamic conditions of the fresh water circulation and the brackish water agitation significantly influence the interface between these two types of water in the coastal areas. Johnston (1983) refers to the salt water-fresh water interface in the Tertiary limestone aquifer in the southeast Atlantic outer-continental shelf (USA). Hydrologic testing in an offshore oil well determined the position

The Origin of Brackish Karst Springs

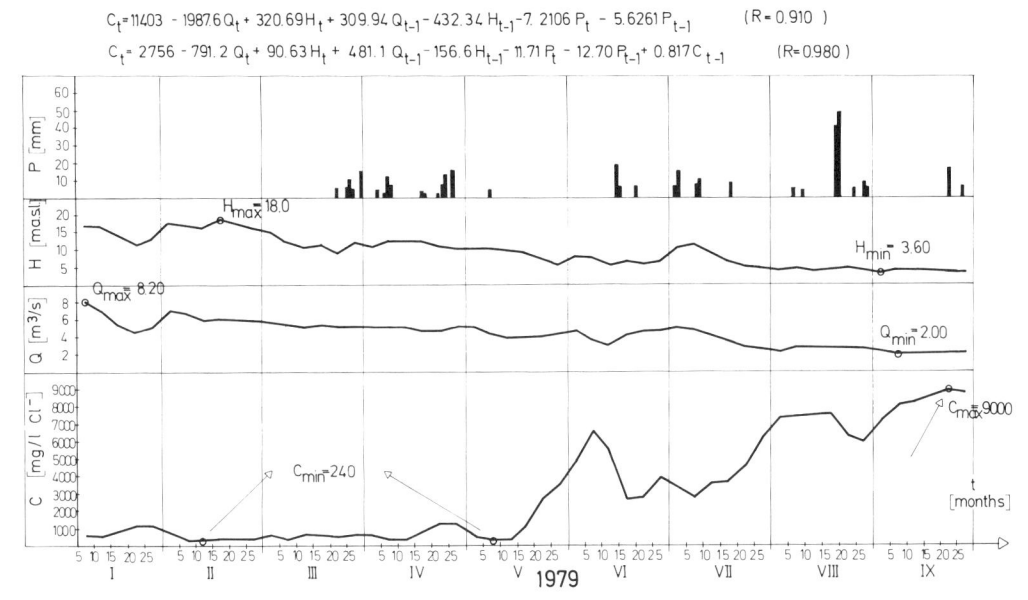

$$C_t = 11403 - 1987.6\,Q_t + 320.69\,H_t + 309.94\,Q_{t-1} - 432.34\,H_{t-1} - 7.2106\,P_t - 5.6261\,P_{t-1} \quad (R = 0.910)$$

$$C_t = 2756 - 791.2\,Q_t + 90.63\,H_t + 481.1\,Q_{t-1} - 156.6\,H_{t-1} - 11.71\,P_t - 12.70\,P_{t-1} + 0.817\,C_{t-1} \quad (R = 0.980)$$

Fig. 4.36. Time series of the groundwater level (H), discharge (Q) and salinity (C) of the Pantan spring (Yugoslavia) and daily precipitation (P) on its catchment area

of the salt water-fresh water interface. Previous drilling established the existence of fresh water 88 km offshore in this area and 335 m below sea level at the coast. A relatively thin transition zone separating fresh water and salt water occurs at a depth of 640 m below sea level. The difference in approximate depth to the fresh water-salt water transition at these two locations suggests an interface with a very slight landward slope. According to Johnston (1983): "Groundwater is constantly discharging seaward beneath the ocean floor by upward seepage through the overlying confining beds in this area. Seaward discharge from a confined aquifer requires that the Hubbert equation (1940) rather than the Ghyben (1889)-Herzberg (1901) hydrostatic relation be applied in estimating the base of freshwater".

Figure 4.36 shows the variations in the salinity of water C in mg l^{-1} at the Pantan Spring (Fig. 4.31) from January to September 1979. The same graph gives the plotted hydrographs of the groundwater level H measured on the piezometer bored in the close vicinity of the spring hinterland and the hydrograph of the spring discharge Q in m^3 s^{-1}. The graph also shows the daily quantity of precipitation P in mm measured by the pluviograph located in the catchment area of the spring. The Pantan Spring is located 300 m off the Adriatic seacoast. Immediately at the spring, there is a small lake ranging from 2.5 to 4 m above sea level. Its water is always salinized so that, in the analyzed period, the minimum measured salinity was 240 and the maximum 9000 mg l^{-1} Cl$^-$. Figure 4.36 shows two multiple linear correlation equations obtained using the theory of the least squares. The dependent variable was, in both cases, the degree of salinity of the Pantan Spring. The coefficient of the linear correlation between the discharge Q and salinity C

is −0.89 and it confirms that a direct connection exists between the spring capacity and its salinity. The coefficient of the linear correlation between the groundwater level H and salinity C is also high and amounts to −0.84. The direct connection between precipitation and salinity is statistically insignificant as the correlation coefficient is only 0.17. This fact can be easily explained since the effect of precipitation is combined with the general condition of the groundwater level and, hence, with the spring discharge. Heavy precipitation in the dry period ($P = 99$ mm, on 19 to 20 August 1979) only slightly influenced the raising of the groundwater level as well as the spring discharge and its salinity. Less intensive rainfall, however, in the wet period of the year significantly influenced the change in the values of these parameters. It can be said that the system displays a certain inertia in its reaction to the outer changes.

Bögli (1980) tried to explain the process of salinization of vruljes and coastal springs using the Bernoulli equation and the hydrodynamic effect. Figure 4.37 A and B gives more detailed explanations of this phenomenon. Fresh water becomes brackish in such a way that the open joints, which take a narrow pass as their point of departure, have a lower pressure than the sea level and thus suck in the sea water (Fig. 4.37 A). Salty sea water can be sucked in when the spring is not submarine, i.e. when it is a coastal spring (Fig. 4.37 B). Lehmann (1932) was the first who explained this phenomenon of salinization of vruljes and coastal springs. It should be noted that this phenomenon is quite rare in natural surroundings, although it is stated in the literature as the only explanation of the salinization of coastal springs and vruljes. There are various types of contamination. According to Kohout (1966) and Breznik (1973 and 1978), brackish karstic springs can be classified in the following ways: (1) springs contaminated due to the hydrodynamic effect; (2) springs contaminated due to the convectional circulation of the saline groundwater induced by geothermal heating; and (3) springs contaminated due to the high density of sea water.

The phenomenon of the springs contaminated due to the hydrodynamic effect has been previously discussed. Gjurašin (1942 and 1943) believes this type of salinization is only of theoretical importance and does not occur in practice.

Springs contaminated due to the convectional circulation of saline groundwater induced by geothermal heating are very rare. They are only found under very deep aquifers as in Florida (USA).

Springs contaminated due to the high density of sea water can be, according to Breznik (1973 and 1978): (1) in the karst aquifer of isotropic permeability and (2) in the karst aquifer of anisotropic permeability.

Karst aquifers of isotropic permeability are similar to aquifers in sediments with intergranular porosity. Thus, the salinization occurs along the entire contact area between the fresh and salt water and not only in the individual karst channels. One of the causes of salinization is the diffusion caused by the mechanical mixing of the fresh and salt water under the influence of the low and high tides, and due to the difference in the velocity of water circulation. Karst aquifers of isotropic permeability are characterized by a great number of springs of small or medium capacity (Breznik 1978).

In the karst aquifer of anisotropic permeability there is a connection between the underground karst fissures and the channels of larger dimensions which con-

The Origin of Brackish Karst Springs

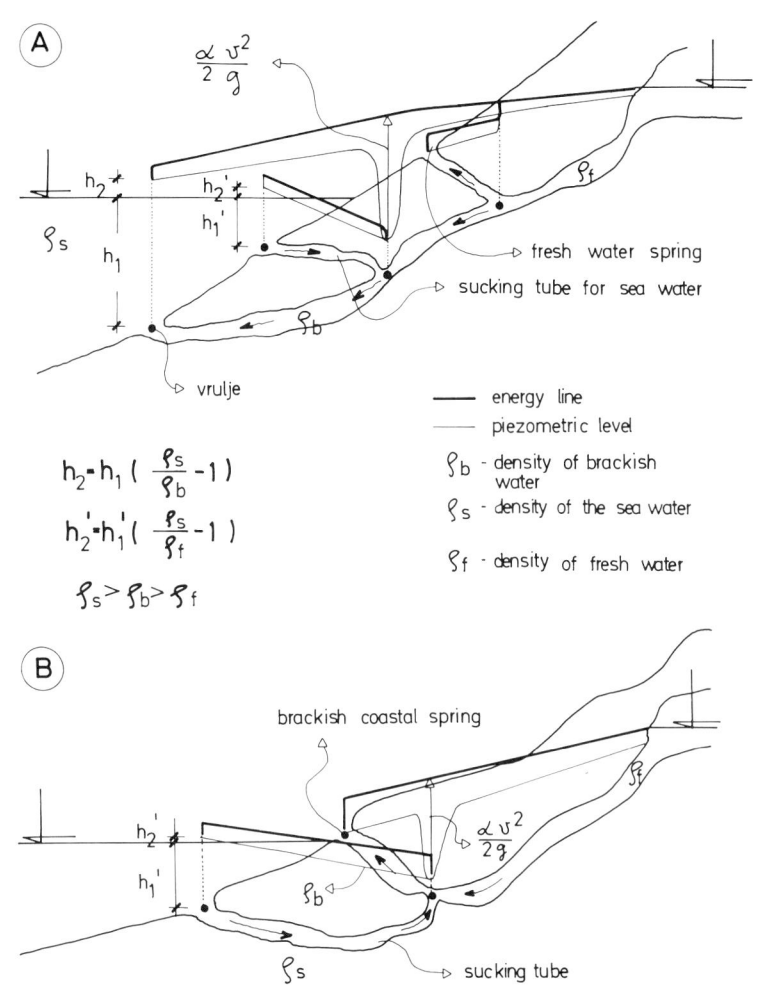

Fig. 4.37 A, B. Functioning of a brackish vrulje and fresh (**A**) and brackish (**B**) coastal springs

centrate into a smaller number of springs with large capacities. Breznik (1973) states that Dinaric karst is a typical example of this type of aquifer. The area covering 17500 km² is drained by only 55 large springs. Gjurašin (1942 and 1943), Kuščer (1950) and Breznik (1973) concluded that there are two types of contamination in the karst aquifer of anisotropic permeability: (1) at the mouth of the submarine springs; (2) inside the karstic mass.

In springs contaminated at the mouth of the submarine spring, the water in the upper part of the inflow channel is fresh and is mixed with the salt water at the mouth of the spring. The sea water penetrates into the current of fresh water due to its greater density. The brackish water formed in this way rises upward due to its density which is less than the density of the sea water. According to Breznik (1973), this phenomenon is similar to the convectional movement of gases above warm objects. Springs of this type are numerous along the karst coasts of Italy and Greece.

The most frequent springs are those contaminated inside the karstic mass. A distinction should be made between salinization of the aquifer with shallow and deep branching. Shallow branching means that the main and the secondary veins are divided at a depth less than 100 m below the sea level. In deep branching, the main and secondary veins branch at a depth of more than 100 m below sea level. The depth of branching depends upon the development of karstification in the geologic past. In both types of springs, their salinity increases as the discharge decreases. In the periods of smaller discharge, the springs with shallow branching can have a sudden increase in salinity. This can be explained by the instantaneous functioning of the secondary vein as a siphon. Springs with deep branching have a more regular variation in salinity than those with shallow branching. This can be explained by a greater distance and depth of branching.

5 Swallow Holes (Ponors)

5.1 Introduction

Ponors are fissures in the karst mass through which the water sinks underground. They play an important role, from the hydrologic-hydrogeologic standpoint, in the water circulation in karst. According to their hydrologic function, ponors can swallow water permanently or can function partly as ponors and partly, i.e. temporarily, as an estavelle. From the morphologic standpoint (Milanović 1981) ponors can include: (1) large pits and caves; (2) large fissures and caverns; (3) sys-

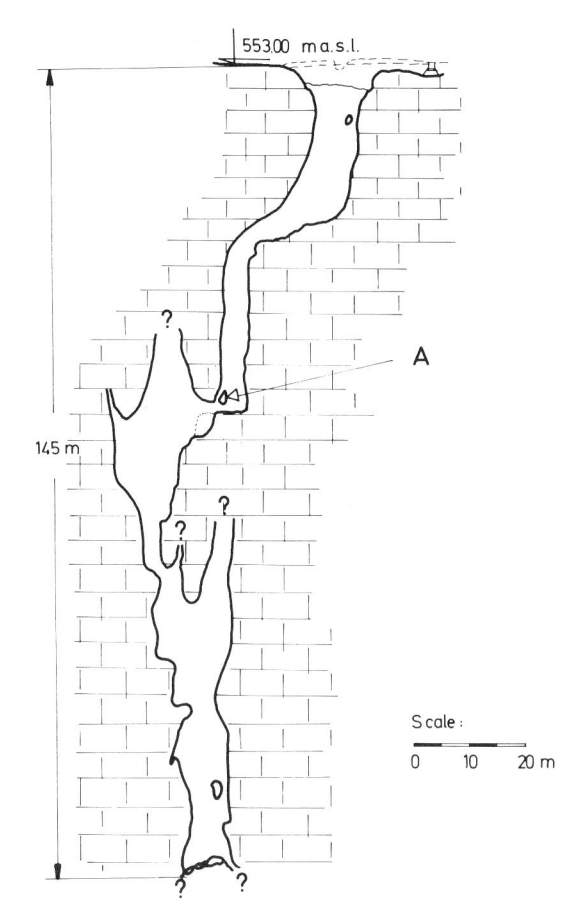

Fig. 5.1. Cross-section through the Tučić Ponor (Yugoslavia)

Fig. 5.2. Photograph of the Ponor mill in the Popovo Polje (Yugoslavia), (taken by Granić)

tems of narrow fissures; (4) alluvial ponors. All underground karst phenomena (jamas, channels and caves) can take over the function of ponors. Jamas most frequently function as ponors and present paths for the direct contact of the surface water with the underground water in the karst mass. Figure 5.1 shows a cross-section of the Tučić Ponor located in the Gračacko Polje (Yugoslavia). It is evidently a vertical jama with a speleologically investigated part 145 m deep, although there are much deeper sections. Figure 5.2 shows a photograph of a ponor in the Trebišnjica Catchment. It is located 30 m above the riverbed of the Trebišnjica and illustrates the size of the flood in the Popovo Polje. The ponor is lined with rock and once it was used as a mill for grinding corn. As a large storage basin was constructed upstream, the polje is not flooded to a high level and, consequently, the ponor and the mill are no longer used. Figure 5.3 shows the entrance to a Jelar Ponor in the Gračacko Polje. The entrance is closed by a grate to prevent its being blocked by logs, plastic material, etc., and to make clearing it possible. The ponor and the Tučić Ponor (Fig. 5.1), with several smaller ponors, represent the only outlet structure from the polje located at 550 m above sea level. Inflow waters from these ponors appear directly in springs in the catchment of the Zrmanja River at 15–25 m above sea level. The grate placed at the entrance to the ponor reveals the attempts of the inhabitants living in those regions to ensure the maximum swallow capacity of the ponors and to reduce, at least partly, the intensity and duration of floods in the polje. Figure 5.4 shows an attempt at closing one ponor-estavelle in the Vrtac Polje (Montenegro, Yugoslavia) in order to form a permanent storage basin. This attempt has not been successful since new ponors

Introduction

Fig. 5.3. Photograph of the entrance to the Jelar Ponor (Yugoslavia), (taken by Granić)

Fig. 5.4. Photograph of unsuccessful sanitization of Ponor-Estavelle in the Vrtac Polje (Montenegro-Yugoslavia), (taken by Bonacci)

Fig. 5.5. Photograph of covering of the Vrtac Polje site by concrete (Montenegro-Yugoslavia), (taken by Bonacci)

were soon formed in its vicinity. The same thing happened when several small ponors were closed. A concrete mat 3–5 cm deep (Fig. 5.5) was made and placed on the surface of the polje above the ponor zone in order to prevent the sinking of the surface water underground. However, numerous new ponors were formed in the vicinity of this area immediately after flooding. These cases point to the specific features of the water circulation in karst which cannot be influenced by the works on the surface of the terrain. The previously mentioned ponors belong to the alluvial type whose origin was explained in detail by Milanović (1981). He also proved that similar surface works cannot ensure the water-tightness of surface storage. In order to emphasize the importance of the ponors, a photograph showing the main ponor in the Cetinjsko Polje is included (Fig. 5.6). The ponor is situated within the town of Cetinje and is used for drainage of 70–90% of all surface waters. The protection of Cetinje and of the karst poljes from floods depends upon the capacity of this ponor. Figure 5.7 shows a photograph of a surface opening of the ponor. The water sinks underground through a large jama. The insufficient capacity of this ponor caused the flooding of Cetinje on 19 and 20 February 1986, as shown in Fig. 5.8.

The catastrophic flooding of Cetinje and to the entire Cetinjsko Polje (Yugoslavia) has drastically emphasized the role of ponors in the process of water circulation in karst. Therefore, the swallow capacity of ponors has to be constantly and carefully considered. The maximum swallow capacity can be ensured by building stone or concrete walls around the ponors or by placing a settling basin

Introduction

Fig. 5.6. Photograph of the Cetinje town and site of the main ponor in it (Yugoslavia), (taken by Bonacci)

with rails which prevent the transportation of fertile soil from the surface into the underground and protect the opening of the ponor from other undesirable materials. At the same time, channels are dug in order to take the surface water to the ponors by the shortest paths. Some of these channels are placed at such a low level as to prevent the flooding of the surrounding area which would take place if they did not exist under natural conditions. Different works are carried out, on the other hand, if the function of the ponors should be decreased or completely eliminated. Such works are necessary before the construction of storage basins in karst. Some unsuccessful attempts at closing ponors have already been discussed (Figs. 5.4 and 5.5). The only reliable technique for decreasing the swallow capacity of ponors is the construction of grouting curtains. It should be stressed, however,

Fig. 5.7. Photograph of entrance to the ponor of the Cetinje Polje (Yugoslavia), (taken by Bonacci)

Fig. 5.8. Photograph of the flood in the Cetinje town caused by insufficient capacity of the ponors (Yugoslavia), (taken by Novaković)

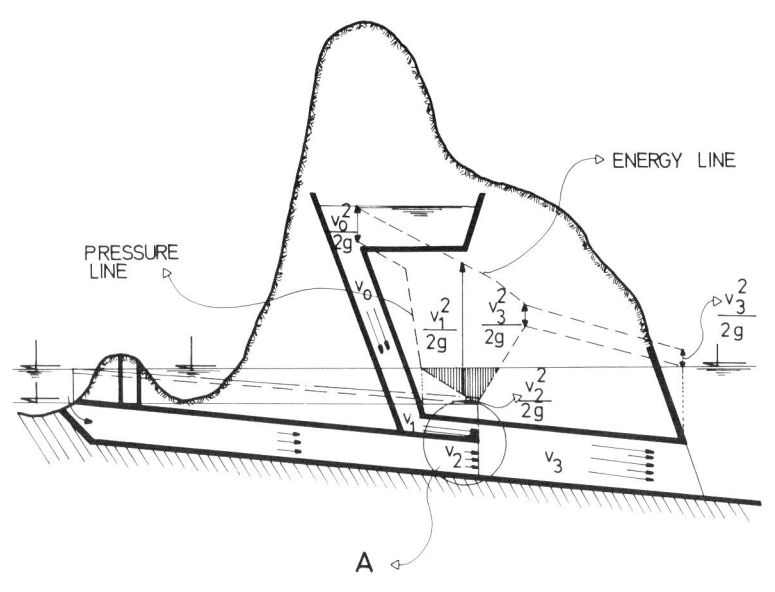

Fig. 5.9. The explanation of the functioning of the sea mills of Agrostali. (Glanz 1965)

that these engineering works are rather expensive and their duration and dimensions are difficult to predict.

The marine (sea) ponors are a very specific and rare phenomenon in karst. Most sea (marine) ponors are of short duration and actually represent the functioning of the vrulja immediately after it dries up. The only permanent sea ponor in the world is the sea mill of Argostoli located on Kephallenia Island in the Ionian Sea (Greece). The functioning of this ponor from the hydraulic standpoint was explained by Glanz (1965) and has been accepted by other hydrologists (Stringfield and LeGrand 1969; Bögli 1980, etc). Glanz (1965) explains the functioning of sea mills by ejector effects (Fig. 5.9) and tries to prove his statement using a physical model. It is hardly likely that the sea ponor of Argostoli actually functions like this. It is physically possible; the model presents an elegant explanation of the phenomenon, but it is difficult to imagine, under natural circumstances, the existence of such a place where all of the karst waters, or at least most of them, are collected, thus resulting in the ejector process. Consequently, the functioning of the permanent sea ponor on Kephallenia Island should be considered as a natural phenomenon not entirely explained.

5.2 Determination of the Swallow Capacity of Ponors

The ponor swallow capacity $_pQ_O$ depends upon the water level H (Fig. 5.10) in the pre-ponor retention only if the flow in the main karst channel is not under pressure. When the flow comes under pressure ($H > H^*$) the discharge curve changes suddenly (point H^*, Q^* on Fig. 5.10). Then the ponor swallow capacity depends exclusively upon the difference ΔH_3 between the level of the pre-ponor retention H and the average level of the spring H_3:

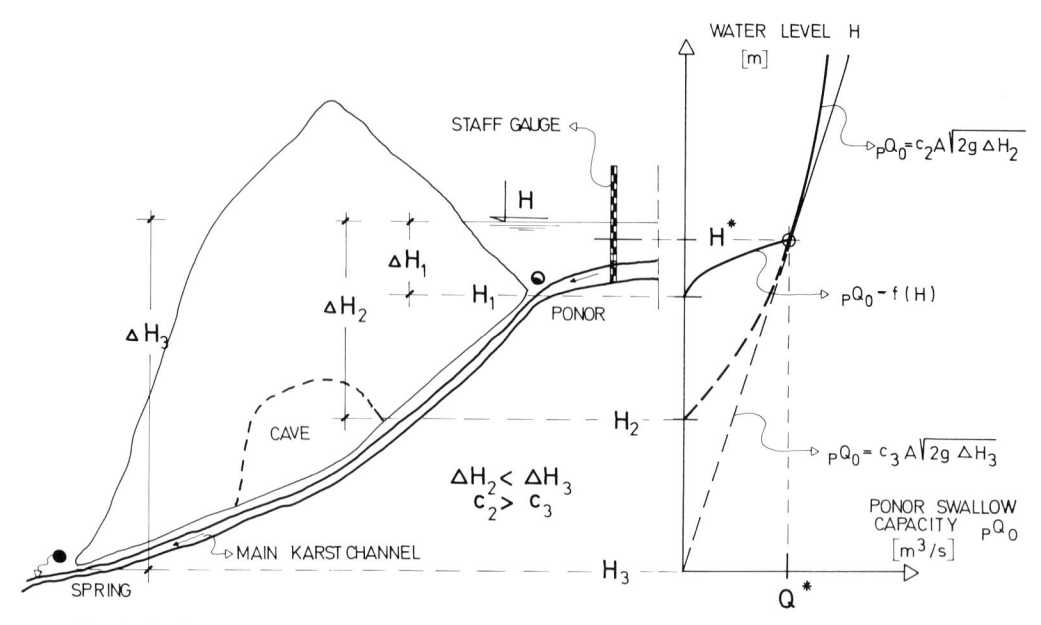

Fig. 5.10. One case of the ponor discharge curve

$$\Delta H_3 = H - H_3, \tag{5.1}$$

and the equation for the ponor discharge curve is:

$$_p Q_O = cA\sqrt{2g\,\Delta H}, \tag{5.2}$$

where c is the discharge coefficient, A is the cross-sectional area of the main channel and g is the gravity acceleration. If there is a large cave system in the karst mass which is never filled with water, i.e. if flow under pressure does not exist all the way up to the spring, then the level ΔH_2 in Eq. (5.2) is lower than ΔH, but the discharge coefficient $c_3 < c_2$. If $\Delta H_3 \gg \Delta H_1$ in cases of flow under pressure (for $H > H^*$), the ponor discharge practically does not depend upon the water level in the pre-ponor retention. This can be seen in Figure 5.11 which shows the curve of the ponor discharges in the Dinaric karst (Yugoslavia).

A general presentation of ponor functioning is given in Figure 5.12. When the groundwater level GWL_1 in the karst mass is higher than the water level in the pre-ponor retention H, the ponor acts as an estavelle. When GWL_2 is lower than H, the ponor swallows water. Then, in that case, Eq. (5.2), for the definition of the ponor swallow capacity, is valid. Measurements should be organized in such a way that the precise value of ΔH can be determined since various situations can occur; the most typical cases are presented in Figure 5.13 (Hajdin and Avdagić 1982).

In order to determine the swallow capacity of independent, large ponors, it is possible to use specially designed measurement devices based upon the principle of measuring velocities or pressure changes at a certain point (Mikulec and

Determination of the Swallow Capacity of Ponors

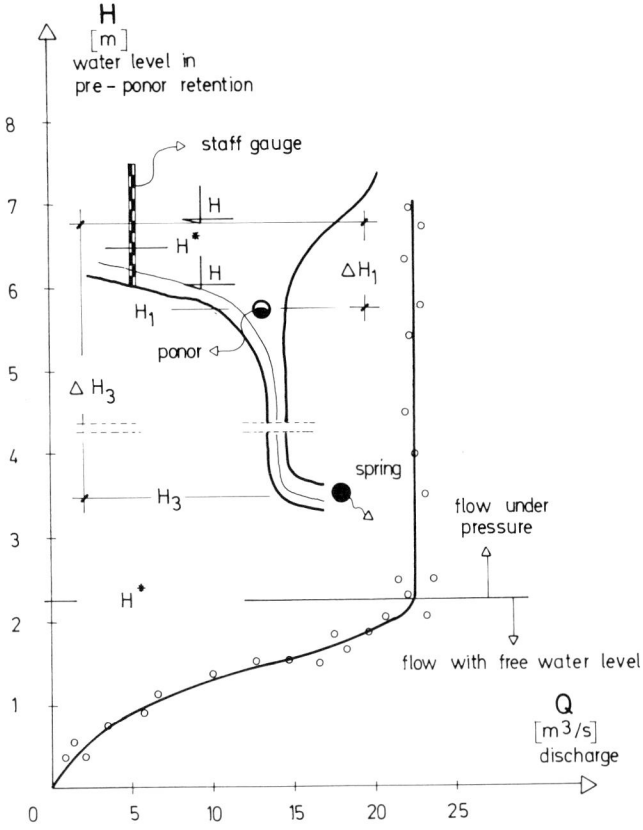

Fig. 5.11. Discharge curve of ponor when $\Delta H_1 \ll \Delta H_3$; Case of ponors in Dinaric karst (Yugoslavia) $\Delta H_{1\max} = 520$ m

Bagarić 1966). In certain situations, when possible, a scale model of a ponor is constructed and the cross-section of the velocity field is defined for different conditions of ponor or estavelle operation. The measurements are carried out continuously with specially designed instruments at one or more points. According to the similarity between the model and the real ponor, the flow to be discharged through the controlled cross-section of the ponor is defined. There are numerous problems related to such measurements, e.g. to protect devices from the floating bodies and to provide the equipment with sufficient energy for long operations, because control over all elements is not possible when the pre-ponor retention is inundated and consequently inaccessible. The dimensions of the equipment are limited to avoid alternations in the conditions of the flow. All previous considerations refer to a case of a large, isolated ponor. These cases are quite simple and not too frequent in water circulation in karst. More frequently, there are numerous ponor zones with several large and a great number of small ponors in one area, and particularly in one polje in karst. In that case, in order to define the swallow capacity of each ponor zone, it is necessary to carry out measure-

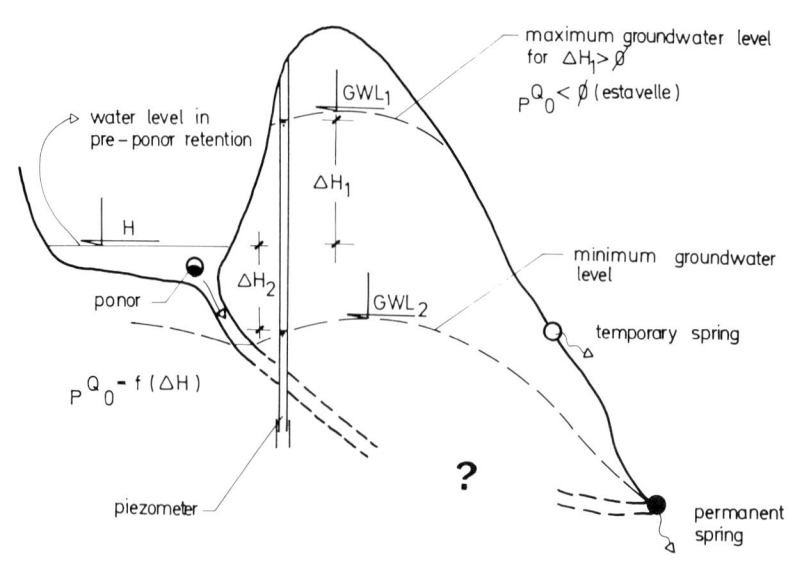

Fig. 5.12. Analysis of ponor capacity in the function of the underground water level

ments of the groundwater levels in that part of the karst mass since they influence the swallow capacity of that ponor zone.

At least one piezometer should be installed for each ponor zone, a group of ponors or a single ponor in order to accurately observe, the changes in the water levels related to the ponors under consideration. The analysis is continued using the budget equation:

$$Q_O - Q_I = \pm (dV/dt), \tag{5.3}$$

where dV is the change in the flood volume in time dt. Inflow Q_I and outflow Q_O discharges consist of numerous elements different for each case which can be determined or measured by one of the known methods. Special attention is paid to the ponor swallow capacity $_pQ_O$ or more precisely, to the outflow from the area effected either though its bottom or sides. Supposing it is a turbulent flow under pressure with the quadratic law of resistance, as proved by various investigations (Hajdin and Advagić 1982; Atkinson 1977; Gale 1984), the following equation for defining the outflow water quantities from ponors $_pQ_O$ is valid:

$$_pQ_O = k \, \Delta H^a, \tag{5.4}$$

where k is the hydrogeologic parameter of the area, and a represents the exponent dependent on the flow regime. Given the turbulent flow with the quadratic law of resistance, a is taken to be 0.5. The value of a can be varied, however, in the course of computations. As there are several ponor zones in the poljes, the above equation is given for each zone since we are trying to obtain the capacity of each ponor or zone using these procedures. On the basis of the hydrogeologic investigations there are six zones in the Buško Blato Polje (Yugoslavia), three zones in the Fatničko Polje (Advagić 1976) and Gračačko Polje (Yugoslavia) and two independent poor zones in the Vrogorac Polje (Yugoslavia). If each zone is

Determination of the Swallow Capacity of Ponors

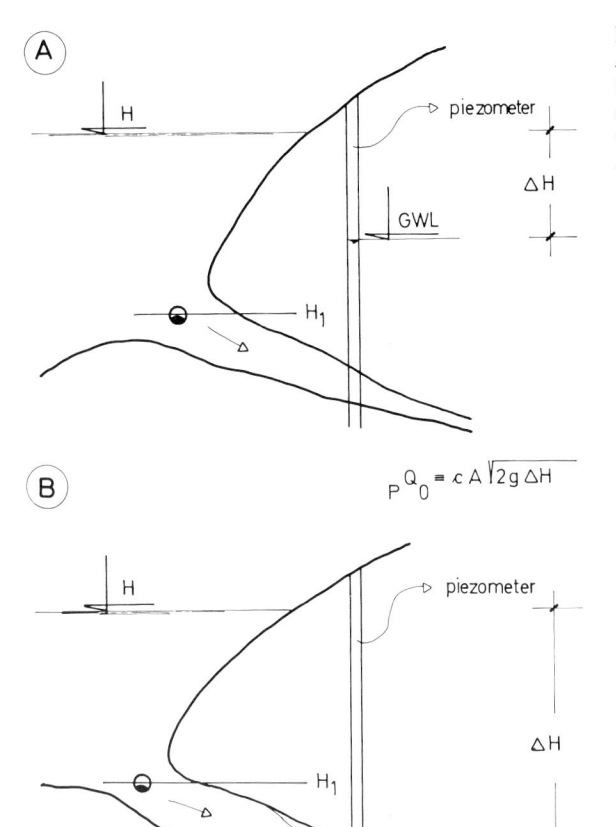

Fig. 5.13 A, B. Determination of ΔH in EW (5.1). (**A**) case when flow in karst pipe is under pressure; (**B**) flow in karst pipe not under pressure. (Hajdin and Avidagić 1982)

$$_pQ_0 = cA\sqrt{2g\Delta H}$$

denoted by index i ($i \in 1, 2, \ldots, n$) and the total number of zones with n, and if each measurement of the inflow and outflow quantities is expressed by j ($j \in 1, 2, \ldots, m$), a system of m equations with n unknowns is obtained:

$$\sum_{i=1}^{n} k_{i1}\Delta H_{i1}^a = Q_{I1} - Q_{O1}^* \pm (\Delta V_1/\Delta t_1);$$

$$\sum_{i=1}^{n} k_{ij}\Delta H_{ij}^a = Q_{Ij} - Q_{Oj}^* \pm (\Delta V_j/\Delta t_j); \quad (5.5)$$

$$\sum_{i=1}^{n} k_{im}\Delta H_{im}^a = Q_{Im} - Q_{Om}^* \pm (\Delta V_m/\Delta t_m);$$

where Q_{Oj}^* is the part of the outflow discharge from the area (polje) which does not include the outflow through ponors, i.e. the ponor swallow capacity $_pQ_O$ (Eq. 5.4). The unknown hydrogeologic parameters of the n zones k_i are obtained

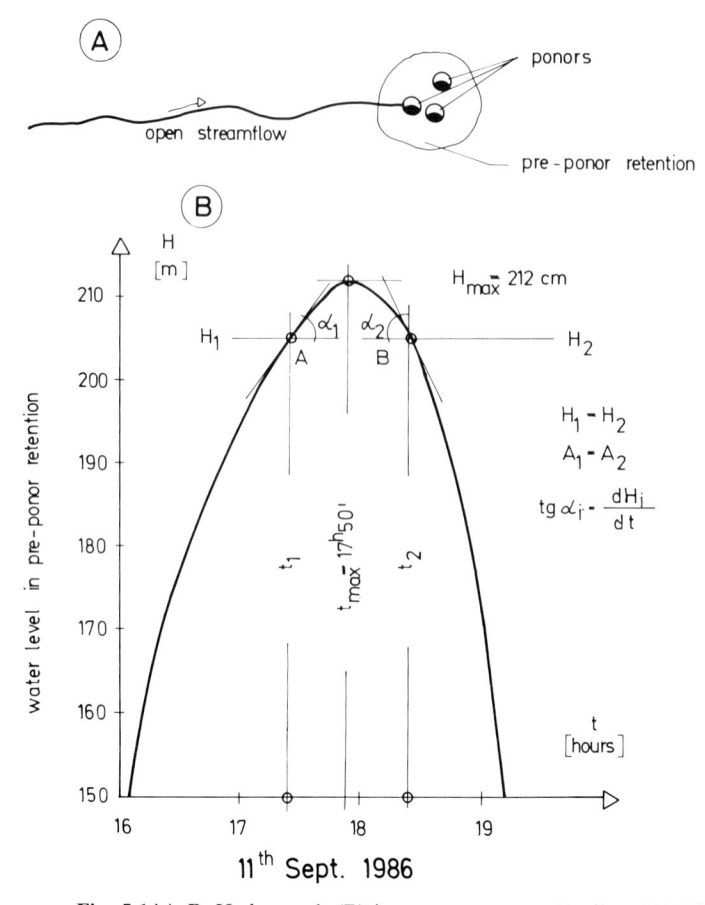

Fig. 5.14 A, B. Hydrograph (**B**) in pre-ponor zone Ponikve (**A**) (Yugoslavia)

by minimizing the sum of the quadratic difference ε_j defined by the following equation:

$$\left(\sum_j Q^*_{Oj} + \sum_i \sum_j k_i H_{ij} - \sum_j Q_{Oj}\right)^2 = \sum_j \varepsilon_j^2 \to \min . \tag{5.6}$$

Accordingly, the number of measurements m should be greater than $2n$ in order to ensure a greater accuracy of the calculation. When applying Eq. (5.5) and the whole procedure, it is necessary to provide constant hydrogeologic parameters of the area k_i. This can be obtained by additional hydrogeologic analyses of the functioning system. Avdagić (1976) developed four systems of equations for the Fatničko Polje (Yugoslavia), valid for different relations between the underground water levels in the flooded polje.

If a river in karst sinks into a ponor zone where the ponor swallow capacity is influenced exclusively by the water level in the pre-ponor retention, the maximum swallow capacity of the ponor can be determined using a relatively simple procedure. The change of the water level in the pre-ponor retention dH/dt is introduced into the budget Eq. (5.3) instead of the volume change in time dV/dt:

Determination of the Swallow Capacity of Ponors

$$(dV/dt) = (dV/dH)(dH/dt) = A(dH/dt) \, . \tag{5.7}$$

The maximum swallow capacity of the ponor zone $_pQ_O$ was obtained by an approximating procedure based on the solution of a system consisting of two equations:

$$Q_{I_1} = {}_pQ_{O_1} + \frac{dH}{dt} A_1 \, ;$$

$$Q_{I_2} = {}_pQ_{O_2} + \frac{dH}{dt} A_2 \, ; \tag{5.8}$$

where Q_I is the inflow into the polje in time t_i ($i \in 1, 2$), $_pQ_{O_i}$ is the swallow capacity of the ponor, ponors or the ponor zone in the same period, A_i is the area of the water surface in the pre-ponor retention and H_i is the water level in the pre-ponor retention. Taking time t_1 and t_2 such that $H_1 = H_2$ and hence that $A_1 = A_2$, it can be supposed that $_pQ_{O_1} = {}_pQ_{O_2}$. This assumption is satisfied if t_1 and t_2 are close and if they are around the maximum water level H_{max} in the pre-ponor retention as presented in Figure 5.14. By plotting tangents on the hydrograph at points $A(t_1, H_1)$ and $B(t_1, H_2)$ the values of dH_1/dt and dH_2/dt are obtained. The inflows the pre-ponor retention should be already known and previously measured quantities. Before selecting the final value, the procedure should be repeated for several different water levels, but all have to be close to the top of the hydrograph (Bonacci 1982). The previously mentioned method is approximate. The accuray of the results depends upon the accuracy of the measurements of the inflow quantities and the water levels in the pre-ponor retention. The method can be applied only if the ponor swallow capacity depends exclusively upon the water level in the pre-ponor retention.

6 Natural Streamflows in Karst

6.1 Interaction Between Groundwater and Water in the Open Streamflows

The hydrological regime of open streamflows in karst depends mostly upon the interaction between the groundwater and surface water. This problem has been dealt with in the previous chapters, particularly with reference to the catchment areas in karst. It has been noted that the groundwater levels in karst vary widely depending upon the effective porosity. Evidently, these variations in the groundwater levels greatly influence the hydrologic regime of the open streamflows. The regulatory influence of karst on the surface circulation depends directly upon the size of the fissures, i.e. the effective porosity, but also on the elevation of the groundwater levels and their change in time. If the catchment area is small and the precipitation intensive, the influence of karst on the surface circulation will not be significant. This fact can be explained since, in that case, there is no adequate volume in the karst mass for the sinking of storm runoff and surface water. The influence of karst differs from one streamflow to another and, therefore, general conclusions should be carefully drawn. Although some open streamflows flow through non-karst terrain, their springs are located in the well-developed karst hinterland. On the other hand, there are streamflows which are formed in a non-karst area, but flow through well-developed karst. The interaction between groundwater and surface water is different, in the two previously mentioned cases as well as in other numerous combinations. The regulatory influence of karst, i.e. the decrease in the high waters and the increase in the low waters, is felt only in the former case. A typical case of such a streamflow, is the Gacka River (Yugoslavia) where the ratio between the minimum, mean and maximum water quantities is $1:4:20$ ($2.5 \text{ m}^3 \text{ s}^{-1}:10 \text{ m}^3 \text{ s}^{-1}:50 \text{ m}^3 \text{ s}^{-1}$) which exhibits a strong regulatory influence of the karst mass. The adjacent Lika River (Yugoslavia) has the ratio $1:100:800$ ($0.25 \text{ m}^3 \text{ s}^{-1}:25 \text{ m}^3 \text{ s}^{-1}:200 \text{ m}^3 \text{ s}^{-1}$). In this case, the karst mass has a weak regulatory influence which is particularly evident in small waters when the river almost dries up. When the open streamflow flows through a karst terrain, the surface water frequently sinks underground and water losses occur along particular sections. This problem will be discussed in detail in Section 6.3. It should be emphasized that the water losses depend upon the groundwater levels, and that they can completely disappear. An illustrative example refers to the four gauging stations on the Krčić River (Bonacci 1985). The essential hydrologic data for these stations (Table 6.1) show how the strong karstification of the catchment affects the average annual discharges which decrease from the spring towards the outlet. When daily discharges are considered, the analysis of the water quantities along the Krčić becomes more complex. The analysis was per-

Interaction Between Groundwater and Water in the Open Streamflows 117

Table 6.1. Hydrologic data along the Krčić River

Gauging station	Distance from the waterfall [m]	Mean annual discharges Q [m^3 s^{-1}]						
		1979	1980	1981	1982	1983	1984	1985
1	9910	7.6	8.1	5.5	5.6	3.6	5.0	3.8
2	8230	7.2	7.7	5.8	5.0	3.1	4.9	3.6
3	2550	6.0	6.2	4.9	4.1	3.0	5.0	2.6
4	20	5.8	6.0	4.8	4.0	3.0	4.8	2.5

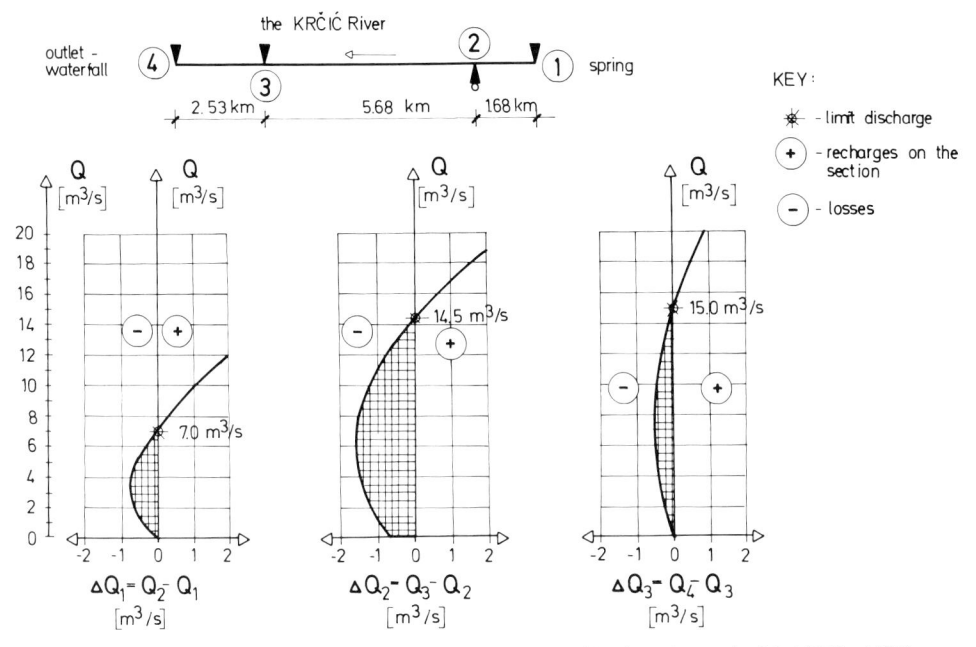

Fig. 6.1. Analysis of losses and recharges along the Krčić River (Yugoslavia) (1979–1983)

formed on a particular section by computing the difference between the daily discharges on each subsequent water gauging station. Figure 6.1 is a graphic presentation of the results obtained by this analysis.

The average daily discharges of the downstream station of the section for day $_jQ_{i+1}$ are plotted on the ordinate, whereas the discharge difference between the downstream and upstream section $_j\Delta Q$ for the same day j is plotted on the abscissa. Consequently, the difference between the daily discharge on the two subsequent stations $_j\Delta Q$ for day j is defined by the following expression:

$$_j\Delta Q_i = {_jQ_{i+1}} - {_jQ_i} . \qquad (6.1)$$

When the value f ΔQ_i is negative, water losses into the karst underground occur on the analyzed section i. On the other hand, the positive sign defines the hydrologically "normal" circulation, i.e. the water of the surface flow comes from underground or at least the losses into the underground are not so significant as to affect "normal" water circulation along the flow. Further analyses show that the first assumption is more probable and can be explained by the fact that during the short periods of normal circulation along the Krčić, the aeration zone of the open streamflow is completely filled by water. Analyzing the result from Figure 6.1, it can be seen that there are constant interactions of the curve of relation $Q_{i+1} - \Delta Q_i$ with the ordinate axis which do not significantly change from one year to another. These intersection points have been named "limit discharges" since they represent very important hydrologic data. The area defined by the curve and the ordinate axis is the negative domain (chequered in Fig. 6.1) and represents an indication of the quantity of losses. The losses are the greatest in the second section, whereas the limit discharges along the Krčić range from 7 m^3 s^{-1} to 15 m^3 s^{-1}. Special attention, from the hydrologic point of view, should be paid to the analysis of limit discharges. They can be explained by the fact that losses due to the karstified catchment either decrease, completely disappear or become smaller than the lateral intercatchment inflow when the discharge is greater than the limit discharge. Its value varies from one section to another. Thus, when the discharge of a particular section of the open flow exceeds the limit quantities, the Krčić catchment does not behave in a way typical of karst, and the discharge increases as the catchment area becomes larger. This fact can be explained by assuming that this happens when the aeration zone becomes saturated. Such situations occur exclusively after heavy rains and are of short duration. Numerous simultaneous measurements carried out on a great number of gauging stations confirm this assumption.

Figure 6.2 gives a graphic presentation of six real combinations of the ratio between the discharge on the downstream profile 2 (on the ordinate) and the difference of the discharges ΔQ on the downstream and upstream profiles (on the abscissa) of an open natural streamflow in karst. Case A refers to streamflows with no losses, i.e. those which are recharged by the groundwater. In this case, the groundwater levels are constantly higher than the water levels in the streamflow. It can also occur that ΔQ is constantly equal to zero (Fig. 6.2B), which means that there is no recharge and no losses. Such a case was observed along a section of the Krka River (Yugoslavia) (Fig. 2.7). The groundwater levels are constantly below the streamflow bottom, but all the karst fissures are clogged by deposits. Since there are no concentrated tributaries along such sections, the water flows along the open riverbed with a constant discharge. Figure 6.2 shows a case when the losses are constantly increased with the increase of the discharge along the downstream or upstream profiles. In that case, the groundwater levels are always below the riverbed bottom. The dashed line represents the second alternative of the same case. Then the discharge on the downstream profile is equal to zero, but on the upstream profile there is an inflow which is entirely lost along the section from profile 1 to 2. In this case, with the increase in the water level of the open streamflow new ponors start functioning so that the losses along the analyzed section of the streamflow constantly increase.

It can be noted that they tend to have a maximum value. The case shown in Figure 6.2D has already been discussed in the example of the Krčić River (Yugoslavia). In this case, the groundwater levels vary. During one time period

Interaction Between Groundwater and Water in the Open Streamflows

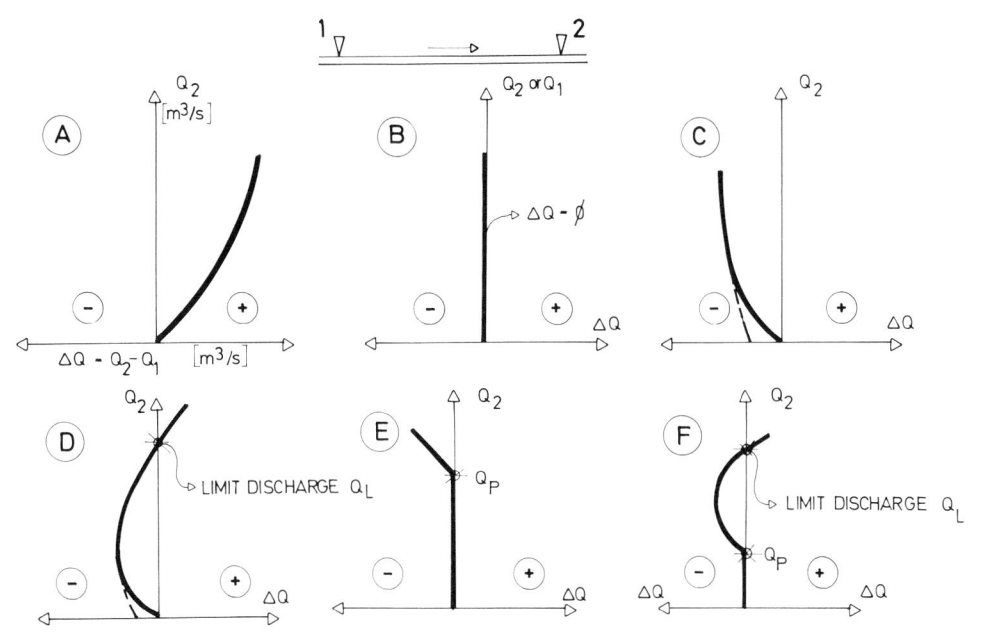

Fig. 6.2 A–F. Real possible combination of discharges relations between upstream 2 and downstream 1 station on the karst open streamflow. (**A**) $Q_2 > Q_1$ no losses; (**B**) $Q_2 = Q_1$ no losses, no recharges; (**C**) $Q_2 < Q_1$ losses constantly increased with the increase of the discharge; (**D**) $Q_2 \gtreqless Q_1$, appearance of "limit discharge" Q_L; (**E**) situation when the ponors are located at an elevation above riverbed bottom; (**F**) combination of case (**B**) and (**D**)

they are below the water level in the open streamflow, and with an increase of the discharge beyond the limit, the groundwater starts to recharge the water quantities in the open streamflow. Figure 6.2E shows a situation where the ponors are located at an elevation above the riverbed bottom. There are no losses until the discharge Q_p occurs due to the colmation of the riverbed bottom and its sides. When this discharge occurs the ponors are activated and the losses are the same as in case C. In this situation, the groundwater levels are constantly below the level of the riverbed bottom. The case shown in Figure 6.2F is a combination of cases B and D. Until discharge Q_p occurs, the riverbed is colmated and there are no losses, although the groundwater levels are below the riverbed bottom. When the limit discharge Q_L occurs, the groundwater levels increase so much that they recharge the open streamflow.

Groundwater flow into open natural streamflows is formed principally under the influence of three factors: (1) climatic; (2) topographic and (3) hydrogeologic structure. According to Koudelin and Karpova (1967), the following six main conclusions can be drawn concerning the effect of karst on the interaction between groundwater and water in open natural streamflows: "(1) Karst leads to an intensification of the groundwater flow. The average long-term, annual maximum and minimum flow moduli (or inches of runoff), and the coefficients of groundwater flow and base flow in karstic regions have markedly greater values, than the regional ones; (2) Karst interrupts the smooth zonal character of the distribution of groundwater flow values over the territory, which ordinarily depend on the climatic latitudinal zonation or vertical zoning in mountain-fold areas; (3) Within karstic regions, wide fluctuations of groundwater flow and surface runoff are observed: areas abundant in water are replaced by arid waterless areas. This is due to the nature of karst and hydrography of karstic regions; (4) Karst leads to redistribution of groundwater flow into rivers within relatively small areas. When studying the groundwater resources of karstic regions by the genetic separation of hydrographs of total streamflow, it is necessary to take sufficiently large "natural basin", within which a full cyle of interrelation (exchange) or surface and groundwaters takes place; (5) The large values of the moduli and coefficients of groundwater flow and base flow in karsted regions are due not only to highly favourable conditions for absorption of precipitation and surface runoff by karstic rocks but are also due to the peculiar features of the karstic water regime noted for its vigorous, turbulent and rapid groundwater movement, relatively direct paths of water flow and well-developed recharge and discharge areas. The average coefficients of groundwater flow for karstic waters are generally considerably higher than those for other types of groundwaters, excluding fissure waters in mountain-fold structures; (6) The type of groundwater flow in karsted massive in many cases (depending on the type of karst) roughly resembles a streamflow regime and has the same phases as the latter, but sometimes may lag behind the streamflow maxima and minima; (7) Karsted rock massives are noted for relatively limited water storage as compared with loose sediments. The average volume of interstices of karst massives for large rock blocks is typically a few percent, whereas the porosity of loose sediments is equal to tens of percent. The values of groundwater flow moduli are conversely related. Accordingly the time of renewal of karstic waters is much briefer than that of waters in loose sediments".

6.2 Hydrologic Regime of Rivers in Karst

In order to define the hydrologic regime of rivers in karst the following standard hydrologic methods can and should be applied: (1) definition of the runoff coefficient; (2) definition of the flow duration curves; (3) analysis of the relationship between the characteristic discharges (e.g. minimum, average and maximum) in a certain time period. In addition to the previously mentioned methods, other hydrologic procedures can be applied. The runoff coefficient is the indicator of the effective infiltration in one area in a certain time period and is defined by the following equation:

$$a = (V_O \pm \Delta V)/V_I ,\qquad(6.2)$$

where V_O is the volume of all the runoff water, ΔV is the change in volume of surface water and groundwater, and V_I is the volume of all waters entering the analyzed area (most frequently a catchment). In order to make it easier to determine the runoff coefficient, a hydrologic year is taken as a unit for the calculation. Thus, the term ΔV can be eliminated from the calculation since it can be supposed to be equal to zero without great error.

Srebrenović (1970) analyzed the runoff coefficients in the karst and non-karst areas in Yugoslavia and defined the following regional equation:

$$a = 0.88 + \frac{215 f}{H} - \frac{420}{H}, \quad (6.3)$$

where H is the precipitation in the analyzed catchment in one hydrologic year, expressed in mm, whereas f is the factor of karst defined by the following equation:

$$f = A_f/A, \quad (6.4)$$

where A is the total area of the analyzed catchment and A_f is the part of the catchment area influenced by karst. Understandably, Eq. (6.3) is the average indicator of the relationship between the values, and the value of a can vary in time and within the analyzed area. Consequently, it is evident that karst influences the increase in the runoff coefficient and this can be numerically proved by Table 6.2. This conclusion is supposed to be valid for karst in general, considering the significant variations occurring from region to region in the values of the runoff coefficient and in the influence of the karst factor on its value. De Vera (1984) presents a contrary statement for the arid to semi-arid conditions in Libya. According to De Vera, the expected runoff in this region is relatively small compared to the catchments without karstification. Generally, the runoff coefficients under the arid and semi-arid conditions in Libya are very low and their yearly values range from 0.0021 to 0.0600 with an average value of 0.021. It can be concluded that in that area the karstification reduces the runoff coefficients of the flood waves of short duration, although it is very difficult to accept the fact that karstification decreases the value of the yearly outflow coefficient. The main influence of karst on the outflow is in the fast sinking of the water underground and the very short retention of water on the surface. These effects can certainly only reduce evaporation, and thus, increase the outflow quantity during a longer time period (e.g. hydrologic year). This influence is stronger if the karst is not covered by soil and vegetation. In the covered karst, without regard to the depth of the covering layer of non-consolidated soil, the outflow coefficient is significantly under a slighter influence of the factor f. Soulios (1984) presents data on the outflow coefficients for three catchments in the karst in Greece. The values of the coefficients vary

Table 6.2. The runoff coefficient depending upon the variation of the karst factor f and yearly precipitation according to the equation defined by Srebrenović (1970)

The karst factor, f	Runoff coefficient a			
	$H = 800$ mm	$H = 1000$ mm	H 1500 mm	$H = 2000$ mm
0	0.355	0.460	0.600	0.670
0.2	0.409	0.503	0.629	0.692
0.4	0.463	0.546	0.657	0.713
0.6	0.516	0.589	0.686	0.734
0.8	0.550	0.632	0.715	0.756
1	0.624	0.675	0.743	0.778

slightly from 0.46 to 0.52. The catchment areas are 9.2 km², 22 km² and 31 km². Since the values of the outflow coefficients are high, it can be supposed that they confirm the previously mentioned conclusion.

The analysis of the discharge duration curves at the gauging stations of karst streamflows yields some significant data on the hydrologic regime of open streamflows and on the influence exerted by karst. In order to compare the data obtained along one streamflow and within one region, the modular coefficients should be used instead of the actual discharge values.

Mimikou and Kaemaki (1985) analyzed the flow duration characteristics of some karstic rivers in Greece. The flow duration curve was regionalized by using the morphoclimatic characteristics (mean annual precipitation, drainage area, hypsometric fall and length of the main river course from the point of division of the drainage basin to a predesignated location) at eleven major flow-measuring sites across the western and northwestern karst regions of Greece. The regionalized regression equations were used to synthesize the flow duration curves at other locations within the hydrologically homogeneous analyzed karst regions. The computations and analyses carried out confirmed the fact that the mean annual precipitation and the drainage area are the most significant factors influencing the formation of the flow duration curves in the Greek karst. The estimated accuracy is satisfactory and varies from 3 to 10%.

Soulios (1985) analyzed the flow duration curves of some karst springs in Greece. The curves were plotted on probability paper. Thus, instead of curves, there were straight, dashed lines. Soulios (1985) explained those breaks in the flow duration curves by the influence of karst on the hydrologic regime of rivers and springs. Figure 6.3 is a schematic presentation of the flow duration curves plotted on chart paper (Fig. 6.3A) and on probability paper (Fig. 6.3B). According to Soulios (1985), the breaks in the curves drawn on probability paper are ascribed to the spilling-over of the karst aquifer towards external outlets or catchments and to the recharging by waters from the surrounding areas. The results of investiga-

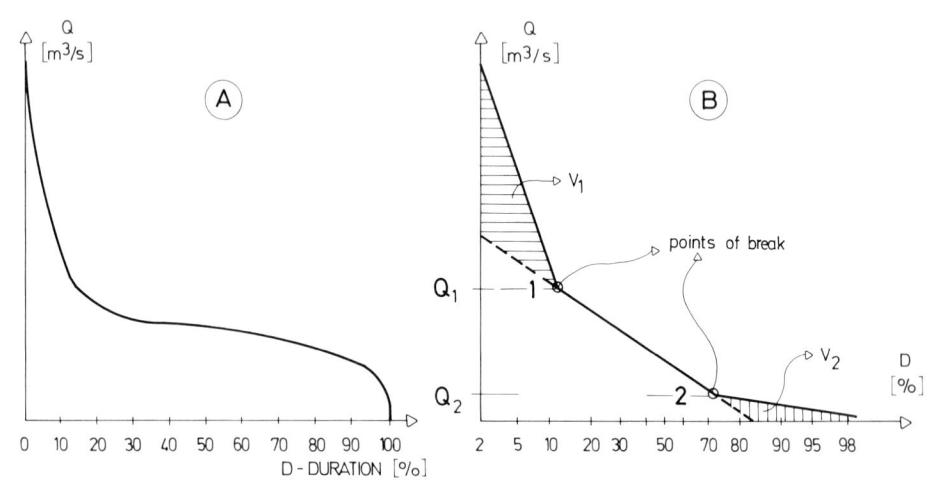

Fig. 6.3 A, B. The flow duration curves on linear (**A**) and probability paper (**B**)

tions carried out on the Krčić River (Figs. 6.1 and 6.2) and conclusions related to the "limit discharge" and water losses on the open streamflow, should be used to complete the given explanation. The breaking points 1 and 2 are shown in Figure 6.3 B. If they actually define the discharge (Q_1), beyond which the catchment is recharged by the adjacent aquifers or beyond which the water losses from the riverbed and aquifer occur (Q_2), then it is possible to approximately calculate volumens V_1 and V_2, expressing the quantities of recharge and losses. The mean flow duration curves for a several year period should be used in order to avoid the influence of the random variations on the accuracy of the conclusions.

The hydrologic regime of the open streamflows in karst differs more or less from the hydrologic regime of the streamflows in the same area which have no surface or underground karst phenomena. Essentially, the influence of karst could be described as regulatory. The value of this influence, however, varies significantly from one region to another and within the same region. The rivers in karst have the long-range amplitude of the annual runoff smoothed down (Gavrilov 1967). The following criteria of the annual flow regulation can be used:

$(Q_l/Q_h) \geqslant 0.6$ High regulation;
$0.6 > (Q_l/Q_h) \geqslant 0.4$ Moderate regulation;
$0.4 > (Q_l/Q_h) \geqslant 0.2$ Slight regulation;
$0.2 > (Q_l/Q_h)$ Negligible regulation.

Q_l represents the average low water, i.e. the mean discharge for the 2 months which have the least streamflow rate (July and August in the moderate belt). Q_h represents the average high water, i.e. the mean discharge for the 2 months which have the highest streamflow rate (most frequently December and January in the moderate belt). The analyses carried out in the Dinaric karst have shown that only a small number of open streamflows have high regulation; most of them have slight regulation. The value of the annual flow regulation should be defined according to the average value over several years. This period must be longer than 10 years in order to eliminate the influence of the random component. These criteria should not be used for those rivers which have water losses along the streamflow caused by sinking through the bottom and riverbed sides.

Markova (1967) analyzed the influence of karst on the rivers of the East European Plains (USSR). The analysis included more than 40 rivers using a series of discharge measurements which were obtained over a period of several years. According to standard hydrologic analyses, it has been concluded that the maximum discharges and the volume of the flood waves of the karst rivers are reduced. The decrease is greater in large, karst-covered areas. The influence of the relationship between surface water and groundwater on the decrease is significant. During low waters, the decrease is greater than during floods. Markova (1967) concluded that the relationship between maximum and minimum discharges is more important in karst than on the rivers flowing through non-karst terrains. This conclusion is contrary to all previously presented conclusions and emphasizes the possibility of great variations in hydrologic values of open streamflows in karst.

It should be stressed that practically every streamflow is specific and should be analyzed separately. One example is a small watershed in the limestone region of eastern Tennessee (USA) (Sodemann and Tysinger 1967). It showed that a change in the forest cover on a watershed in karst was associated with a change in the water yield. A increase in the forest cover over a long period (30–50 years), resulted in a decrease in the water yield. An increase in the water yield was noticed after the cutting of the forest.

Balkov (1967) concluded, according to the hydrologic analyses of numerous karst streamflows in the USSR that karst affects the redistribution of the discharges during the year by increasing the moduli of minimum discharges, where q_{min} is expressed in m^3 s^{-1} km^2. The increase in the moduli of the minimum discharges of karst rivers is inversely proportional to the catchment area. Balkov (1967) stated that each wide karst region has a different limit of maximum catchment area A_{max}. The influence of karst on the minimum discharge is not noticed for the catchments beyond that value within a greater area. The influence of karst on the redistribution of the discharges during the year decreases as the humidity of the catchment increases. The humidity refers primarily to the value of the modulus of the mean yearly discharges and then to the yearly precipitation and groundwater levels forming the recharge of the river. The previously presented conclusions are valid only on karst rivers flowing through plains which are recharged by groundwater during the year.

6.3 Water Losses Along the Open Streamflows in Karst

When discussing problems of measuring and establishing water losses along river channels in karst, reference is made primarily to upstream and middle reaches of the stream which are not located in typical and easily discernible ponor zones, and where the disappearance takes place through barely perceptible small fissures, most frequently at the bottom, and rarely at the sides of open streams (Bonacci 1981). To determine these losses the extremely expensive and sensitive process of "simultaneous discharge measurements" must be carried out (Bonacci and Perger 1970). On the basis of these measurements, far-reaching and often very expensive engineering decisions are made on the need for grout injections, reservoir locations or other remedial measures.

In the first place, field works should be organized so that as many flow discharge measurements as possible can be carried out in the period of stagnant water levels. It is necessary, from the statistical aspect, to make not less than n such measurements under identical hydrologic conditions (Bonacci 1979). The amount of n is defined by the expression:

$$n = \frac{u^2 \sigma^2}{\varepsilon^2}, \tag{6.5}$$

where u is a standardized variable of normal unit distribution, σ is the standard deviation of the basic group and ε is the given scope of interval. Table 6.3 includes a series of values of the sample n, on the condition that a reliability of 95.45% is required to which $u = 2$ corresponds. Analysis of the numbers given in Table 6.3 shows that the minimum number of necessary measurements ranges from 4 to 16 for $\varepsilon = 0.2$ m^3 s^{-1} and 0.2 m^3 s$^{-1} \leqslant \sigma \leqslant 0.4$ m^3 s^{-1}, which is in fact very difficult to achieve.

Table 6.3. Required number of measurements n with a reliability of 95.45% ($u = 2$)

ε [m³ s⁻¹]	σ [m³ s⁻¹]					
	0.1	0.2	0.3	0.4	0.5	0.6
0.10	4	16	36	72	100	400
0.20	1	4	8	16	25	100
0.30	1	2	4	8	11	44

The most favourable situation for discharge measurement, and for establishing water losses along sections of open streamflows in karst, occurs when water stage levels remain the same for a period of several days. In such cases, all measurements can be carrried out under identical conditions. The work on the site should be organized in such a way as to carry out the greatest number of discharge measurements while the water table levels are stable. In practice, the minimum number of measurements ranges from eight to ten. With this type of measurement it is very interesting to study the influence of the gauging team and its equipment on the variation of measurement results. The computation is completed using an analysis of variance and applying Latin square schemes. The work scheme with all the necessary explanations is given in Figure 6.4 (Bonacci 1982). To carry out such an analysis, the measuring instruments and the teams need to be changed for different cross-sections according to a previously established plan. One such possibility is given in the matrix of the Latin squares in Figure 6.4. Although fulfilling this condition calls for additional time, results can point to significant factors influencing measurements. Further explanation is provided by the mathematical model given in Eq. (6.5), which represents the essence of variance analysis using Latin squares:

$$\Delta Q_{ijk} = \bar{Q} + \Delta Q_i + \Delta Q_j + \Delta Q_k + \varepsilon , \qquad (6.6)$$

where ΔQ_i is the influence of the team on the measurement (discharge Q_{ijk}), ΔQ_j is the influence of measuring instruments, ΔQ_k is the influence of the measurement cross-section, ε is the influence of uncontrolled factors (most often taken as a random unexplained part) and $\bar{Q}$ is the expectation (approximated by the arithmetic mean). This scheme can also be combined with an analysis of the influence of the water level change on the variations in measurements. This can be achieved by applying the Graeco-Latin squares method in the analysis of variance as exemplified in Figure 6.4. The number of measurements remains the same as in Latin squares.

Figure 6.5 shows the results of simultaneous measurements carried out on seven cross-sections of the Gornja Dobra River. The discharge was measured five times on each profile ($n = 5$), and the mean discharge on the inflow profile was 2.20 m³ s⁻¹. The measurements encompass a section of the river 17 km long. The individual sections had different lengths ranging from a minimum length of 875 m to a maximum length of 4875 m. The maximum losses occurrred along the last section between profiles 6 and 7 and amounted to 357 l s⁻¹. Along the fourth section (between profiles 4 and 5) there were no losses; the discharge, how-

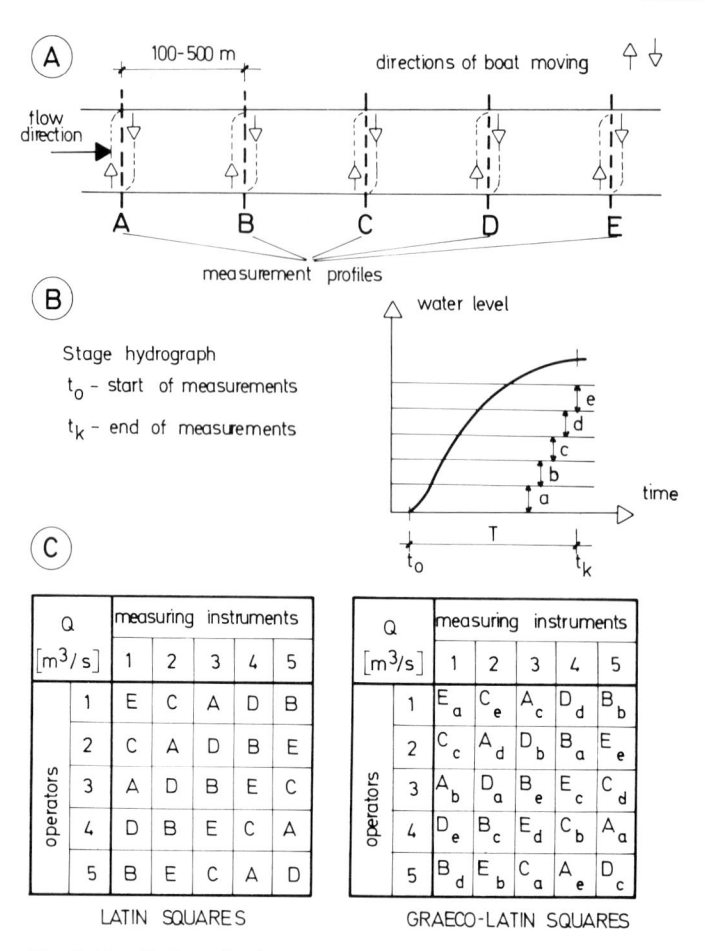

Fig. 6.4A–C. Organization scheme for simultaneous measurements for defining water losses along open streamflow. **A** Situation with measurements profiles; **B** stage hydrographs; **C** Latin and Greco-Latin squares

ever, increased by 127 l s^{-1}. The presented example confirms the existing variations in the behaviour of the losses and inflows along the natural streamflow in karst. All measurements were carried out simultaneously applying the "velocity-area" method. In this actual case there were ten teams engaged in this field work. Seven teams carried out the measurements on the profile of the Gornja Dobra River and three teams controlled the discharge quantities of the small tributaries. The inflows and losses between profile i, $i+1$ were defined by $Q_{i,i+1}$ and were estimated along the intervals using the following equation:

$$Q_{i,i+1} = \bar{Q}_{i+1} - \bar{Q}_i \pm s_{\Delta Q_{i,i+1}}, \tag{6.7}$$

where Q_i is the mean discharge measured on profile i, and $s_{\Delta Q_{i,i+1}}$ is the standard error of the value $Q_{i,i+1}$ defined by the equation:

$$s_{\Delta Q_{i,i+1}} = \sqrt{s_{Q_i}^2 + s_{Q_{i+1}}^2}, \tag{6.8}$$

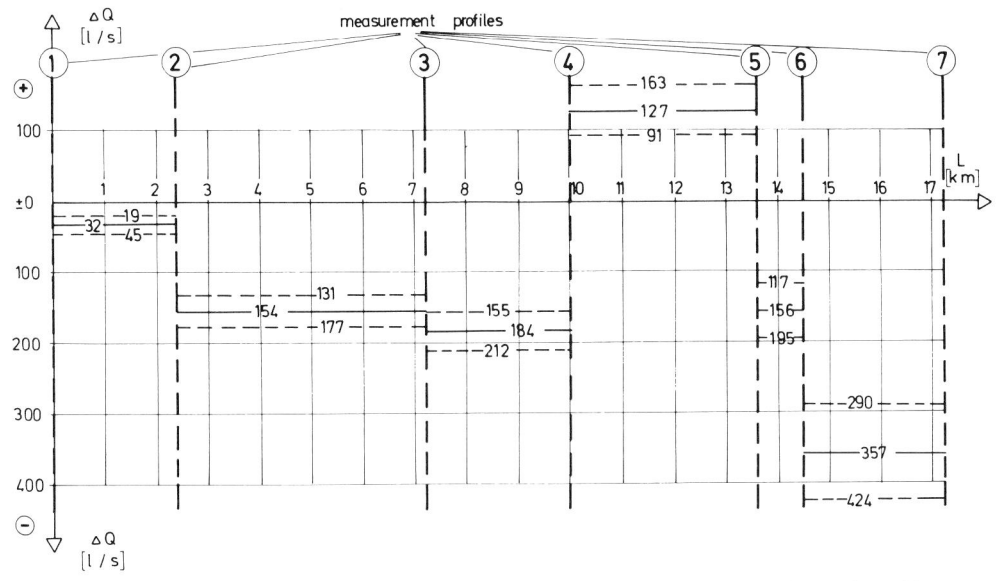

Water Losses Along the Open Streamflows in Karst

Fig. 6.5. Results of simultaneous discharge measurements on the Gornja Dobra River (Yugoslavia)

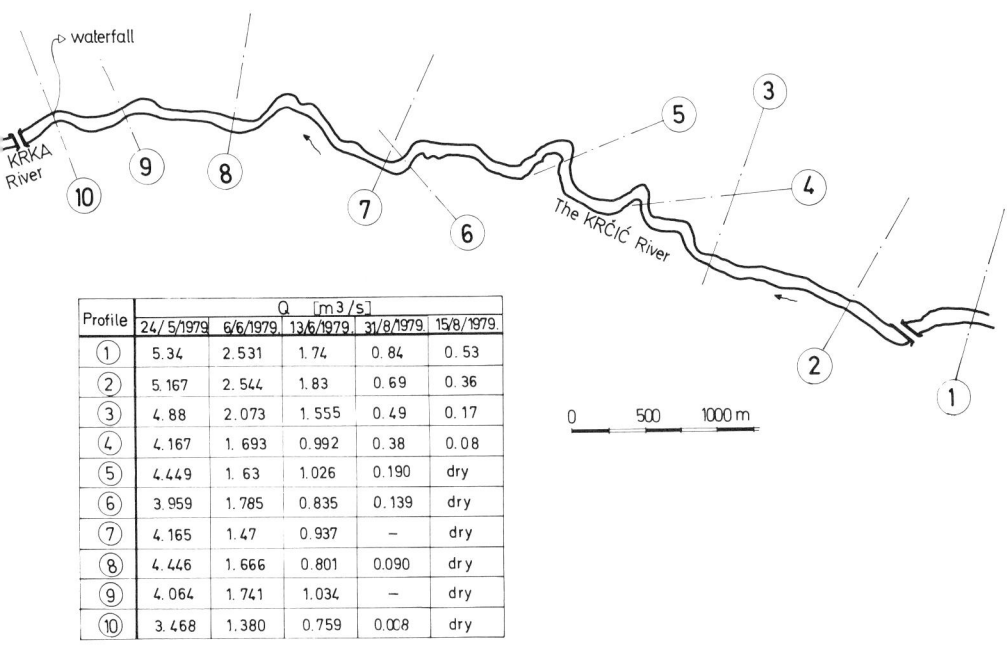

Fig. 6.6. Situation of a part of the Krčić River with results of simultaneous discharge measurements on ten profiles (Yugoslavia)

Profile	24/5/1979	6/6/1979	13/6/1979	31/8/1979	15/8/1979
1	5.34	2.531	1.74	0.84	0.53
2	5.167	2.544	1.83	0.69	0.36
3	4.88	2.073	1.555	0.49	0.17
4	4.167	1.693	0.992	0.38	0.08
5	4.449	1.63	1.026	0.190	dry
6	3.959	1.785	0.835	0.139	dry
7	4.165	1.47	0.937	–	dry
8	4.446	1.666	0.801	0.090	dry
9	4.064	1.741	1.034	–	dry
10	3.468	1.380	0.759	0.008	dry

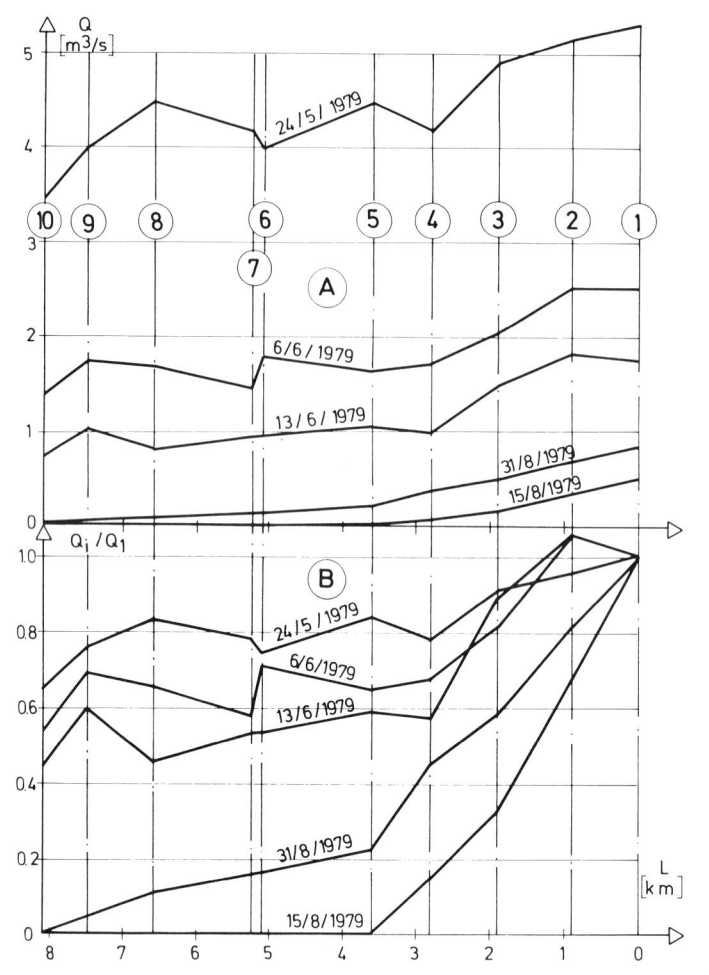

Fig. 6.7 A, B. Graphical presentation of simultaneous discharge measurements taken out along the Krčić River (Yugoslavia) in absolute discharges Q_i (**A**) and relative discharges Q_i/Q_1 (**B**)

with the standard deviation

$$s_{Q_i} = \sqrt{\sum_j^n (Q_{ij} - \bar{Q}_i)/(n-1)} \ . \tag{6.9}$$

The statistical significance of the difference between discharges $\Delta Q_{i,i+1}$, measured on the profiles i, $i+1$ can be proved by applying the t-test and the F-test. It has been proved by such testing that the difference between the discharges measured on profiles 1 and 2 is not statistically significant, and thus, is supposed to be a random value. For all other sections the differences between the discharges $\Delta Q_{i,i+1}$ are statistically significant.

Figure 6.6 illustrates the situation downstream in the Krčić River (Yugoslavia) when five simultaneous discharge measurements were carried out on ten charac-

Water Losses Along the Open Streamflows in Karst

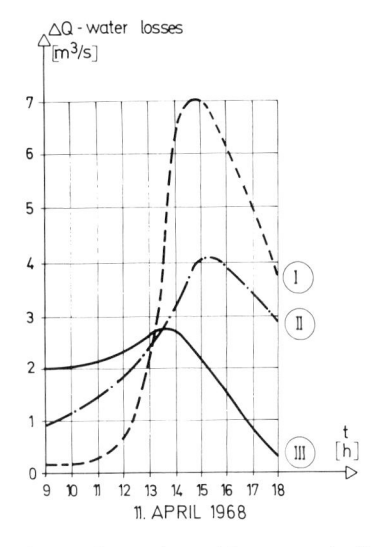

Fig. 6.8 A–C. Measurement results for water losses along a karst river with nonsteady flows. **A** Scheme indicating section and measurement profiles; **B** discharge curve for profile 2; **C** diagrams of sinking in sections

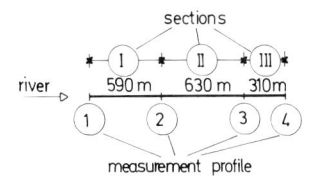

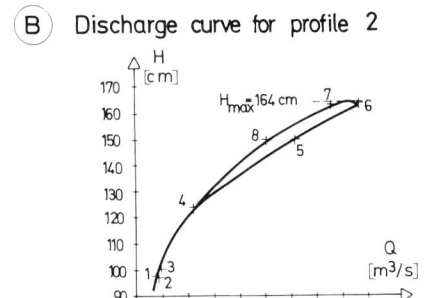

teristic profiles. Also shown are the results of the mean discharges. The graphic presentations of these measurements are given in Figure 6.7A and B. Figure 6.7A shows the results in absolute amounts, and Figure 6.7B in relative values. Each value was divided by the inflow discharge measured on the first profile. It can be seen that generally the discharge decreases as the river approaches its estuary and while in the spring zone the discharge varies from 0.7 to 0.9 m³ s⁻¹, at the estuary the river is completely dry. All measurements were carried out when the groundwater levels were below the levels in the open streamflow.

It is not always possible to carry out discharge measurements under steady flow conditions even though it is preferred. The determination of water losses demands a very high accuracy and unstable channels whose section varies over a period of time should be avoided at all costs. An example of measurements obtained on a river in Yugoslav karst is given in Figure 6.8. The computation of swallow capacity per section was effected on the basis of a budget equation expressed as follows:

$$V_I = V_O + G \pm R , \qquad (6.10)$$

where V_I is the water volume at the inflow cross-section, V_O is the water volume at the outflow cross-section, G is the volume of losses along a section and R is the volume of water retention along that section. Since the inflow and outflow quantities are approximately the same, the entire computation is carried out with successive approximations, with the requirement that during measurement period T from the beginning of t to the end of t_k, conditions of conservation should be

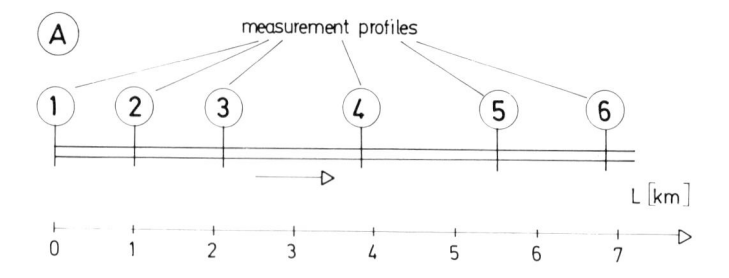

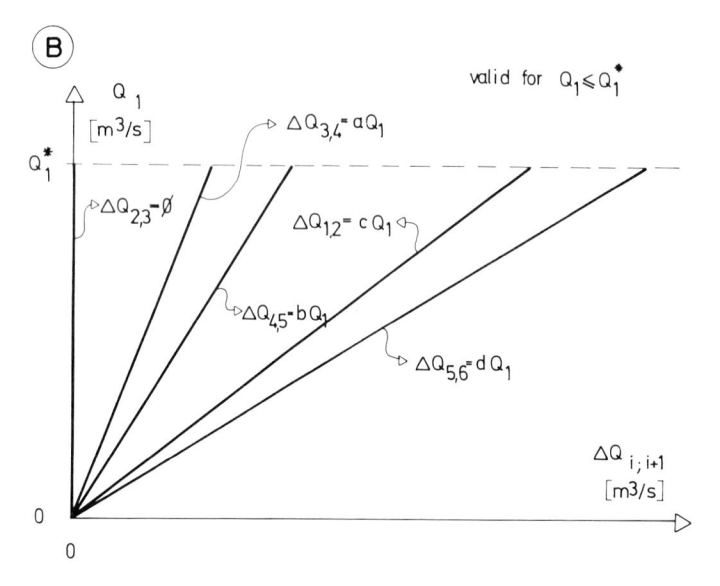

Fig. 6.9 A, B. Linear dependence of losses on the section $\Delta Q_{i,i+1}$ on the inflowing discharge Q_i. **A** Situation with measuring profiles; **B** discharge curves for losses

fulfilled, i.e. $_TV_I$, the water volume at the inflow cross-section during the entire time period T should be equal to the sum of $_TG_I - I(1, 2, \ldots, n)$, the volume of water sinkage into the karst underground in the I-th section during the entire time period T. It is supposed that the retention volume does not change, and the computation begins and ends with the same retention. During the period when storage is decreasing, a budget equation for each time increment Δt (from t_1 to t_2) within time T must be satisfied. The main disadvantage of such measurements is that results and conclusions obtained in this way are often valid only for a specific situation, whereas they cannot be directly applied to different conditions of time and underground water in karst. For this reason, it is necessary to measure levels of underground water near the riverbed on lines hydrogeologically connected with the analyzed open watercourses of ponor zones, since water levels can have a significant influence upon the swallow capacity of the karst underground.

After establishing the water losses in a karst river it is necessary to try to define the quantity of those losses according to the discharge at the inflow profile. This dependence can be linear as in the case shown in Figure 6.9. This shows that the

Water Losses Along the Open Streamflows in Karst

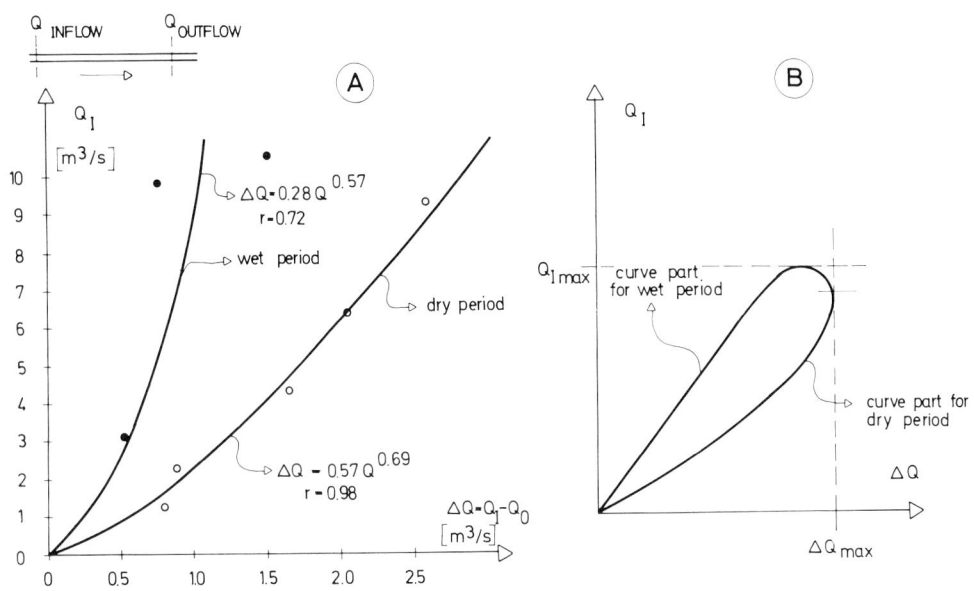

Fig. 6.10. A Dependence of losses ΔQ on one section of the open karst streamflow Cetina (Yugoslavia) on general wetness and inflowing discharge Q_i. **B** General case

dependence varies from one section to another and that it is valid only for a certain area of the inflow profile discharge. Most often this discharge amplitude varies from zero up to a certain limit value Q^* beyond which the losses are influenced by the groundwater levels as well. The described case represents the simplest form of dependence which is frequent in the Dinaric karst of Yugoslavia. There are more complex forms of dependence and in order to study them, the groundwater levels along the river should be available. Most frequently such data are not available and the problem can be solved only approximatively. An example of such a case is given in Figure 6.10. It is a section of the Cetina River (Yugoslavia) where the losses are significant and range from 0.5 to 3 m^3 s^{-1}. In this case, the losses are neither linear nor uniform in the function of the inflow discharge Q_I; however, they depend upon the humidity which can be best defined by the groundwater levels. Since such data are not available, it has been supposed that the dependence of the losses on the inflow discharge for the cold and wet period is different from the dependence in the warm and dry period. The general form of that dependence is illustrated in Figure 6.10A. Although only five simultaneous measurements were carried out in the dry period and three measurements in the wet period, the presented assumption was confirmed by the water level measurements and the discharge curves of the outflow and inflow profiles. It should be stressed that the inflow and outflow profiles have a staff gauge and a discharge curve for each of them is available.

A special problem for open streamflows in karst occurs when there is no flow in the riverbed, i.e. during the drying up period. This phenomenon occurs quite frequently in the Dinaric karst (Yugoslavia) even in those rivers with a catchment

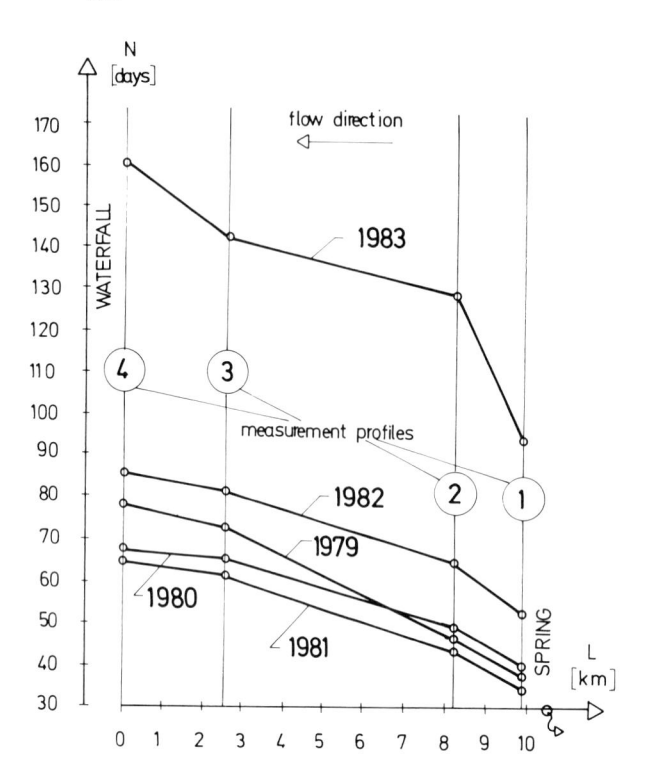

Fig. 6.11. Number of days (*N*) with no flow along the Krčić River (Yugoslavia)

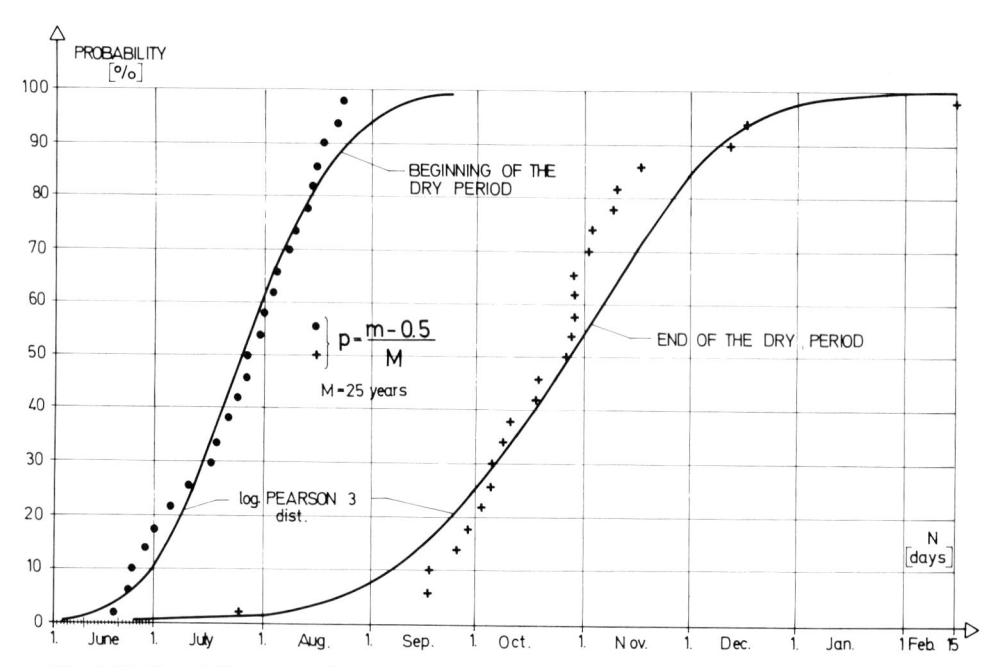

Fig. 6.12. Probability curves for the dates of the beginning and the end of dry period of the Krčić River at Station 3 (1959–1983) (Yugoslavia)

Water Losses Along the Open Streamflows in Karst 133

Table 6.4. Number of drying up days N at two gauging stations on the Zrmanja River (Yugoslavia)

| Ordinal No. | Year | N Days at gauging stations ||
		1[a]	2
1.	1953	–	121
2.	1954	–	102
3.	1955	–	110
4.	1956	–	93
5.	1957	–	34
6.	1958	–	79
7.	1959	0	75
8.	1960	29	104
9.	1961	51	98
10.	1962	108	140
11.	1963	0	28
12.	1964	0	61
13.	1965	0	25
14.	1966	0	26
15.	1967	30	52
16.	1968	0	23
17.	1969	0	26
18.	1970	88	146
19.	1971	77	139
20.	1972	0	47
21.	1973	18	95
22.	1974	0	20
23.	1975	0	71
24.	1976	–	14
25.	1977	–	70
26.	1978	–	0
27.	1979	–	3
28.	1980	0	26
29.	1981	0	56
30.	1982	0	58
31.	1983	135	184
	$\bar{N}$	25.5	68.3

[a] – Indicates no available data.

area covering more than 500 km². Identical situations are found in the open streamflows of the karst rivers in Turkey, Greece, Libya, Spain, etc. The hydrologic characteristics of the drying up phenomenon can be illustrated by the example of the Krčić River (Yugoslavia). Figure 6.11 gives the number of drying up days in the period between 1979–1983. The drying begins downstream and gradually reaches the spring. Measurements and analyses show that the Krčić River dries up with a yearly average for a particular station in the following way: Gauging Station 1, 46 days; Station 2, 57 days; Station 3, 80 days (min. 19–max. 143); Station 4 (outlet-waterfall), 90 days. The dry period begins most often at the beginning or middle of July and ends by the middle or end of September. The drying up of the Krčić was recorded only twice in a winter period. Figure 6.12 shows the

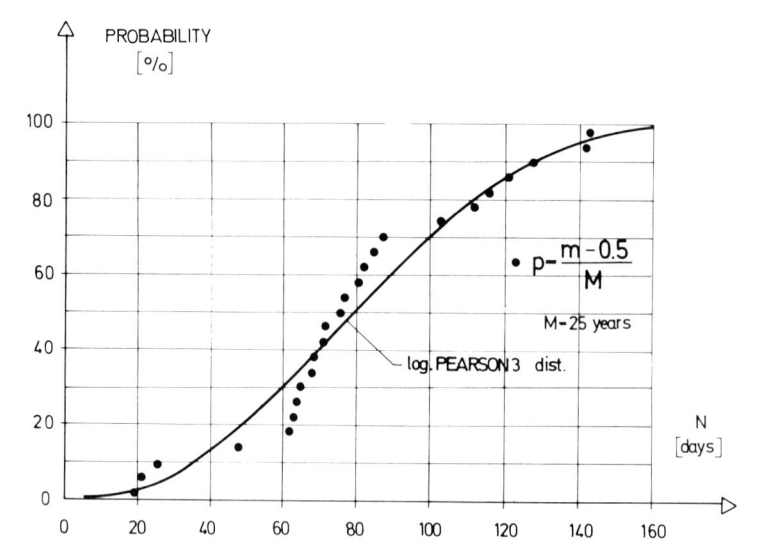

Fig. 6.13. Probability curve for number of days N with no flow on the Krčić River at Station 3 (1959–1983) (Yugoslavia)

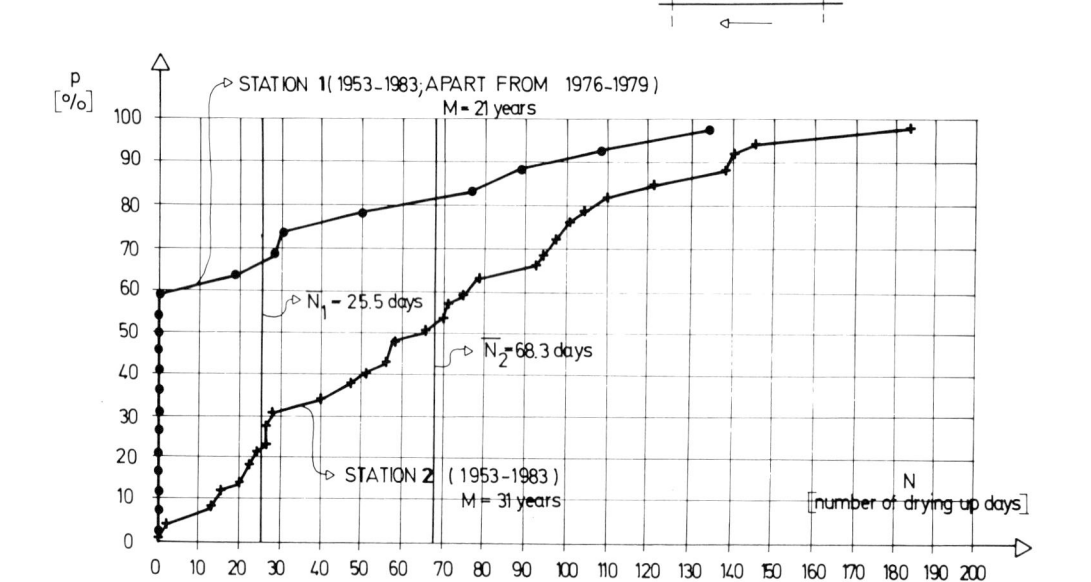

Fig. 6.14. Empirical probability curves for the number of drying up days on two stations at the Zrmanja River (Yugoslavia)

probability curves for the beginning and the end of the dry period of the Krčić at Station 3 between 1959–1983. The beginning of the period covers a relatively short time span, while the end of the period covers a very wide time span from September to February. The end of the dry period depends upon the first heavy autumn rains which are sometimes late and start falling in winter. Figure 6.13 shows a probability curve for the number of drying up days during one year at Gauging Station 3 in the period between 1959–1983.

The Zrmanja River (Yugoslavia) dries up on two profiles. Table 6.4 shows the number of drying up days N, in one year at Gauging Stations 1 and 2. It is evident that on profile 1 the Zrmanja dries up only in 40% of the years, and then the drying up period lasts a long time. On the average, the Zrmanja dries up 26 days on the profile (if all the years are taken into account) or in 67 days if only the years when the drying up occurs are considered. The year 1983 was especially dry and there was no surface flow for 135 days, and on profile 2 for 184 days, i.e. approximately 6 months. On profile 2 there was no drying up only one year (1978), and on the average the Zrmanja has no surface flow on that section for 68 days a year. The drying up is very rare in the winter period. It occurred only twice. In order to find out where this water reappears, and to extensively study the dynamic aspect of this phenomenon, a greater number of surface measurement points should be established along the Zrmanja. It would be important, however, to have a network of well-studied piezometers for measuring the groundwater levels. Figure 6.14 gives the curves of the empirical probability for the distribution of the number of drying up days in the open streamflow of the Zrmanja on profiles 1 and 2. It is a graphic representation of the data given in Table 6.4.

7 Hydrologic Budget for the Poljes in Karst

The poljes in karst have already been dealt with in Chapter 2 as a morphologic and surface phenomenon. As previously stressed, from the hydrologic standpoint, poljes represent a subsystem linked to the surrounding karst mass, other poljes or river valleys on higher horizons and to those poljes or river valleys on lower horizons. The preceding chapters discussed a few examples of the influence of the surface waters and groundwaters of higher and lower horizons under hydrologic conditions in the analyzed polje. When defining the water budget in the poljes difficulties arise in the flooding periods. In such situations it is very difficult, even impossible, to control the water inflow. Furthermore, the water then flows into the polje through a series of temporary springs and the estavelle function as springs. Figure 7.1 shows a probability curve for the dates of the occurrence of the maximum flood level in the Konavosko Polje near Dubrovnik (Yugoslavia). In addition to this curve, it is necessary to define the probability curves for the beginning and ending dates of the flood, as well as for the flood duration in the poljes. Flood analysis should be performed according to a schematic representation of the inflow and outflow hydrographs, i.e. according to the analysis of floods in the pre-ponor retention. Figure 7.2 A gives a graphic presentation of the inflow and outflow discharge hydrographs. It is the simplest case discussed by Ristić (1976). Figure 7.2 B presents the water level hydrograph in polje $H = f(t)$

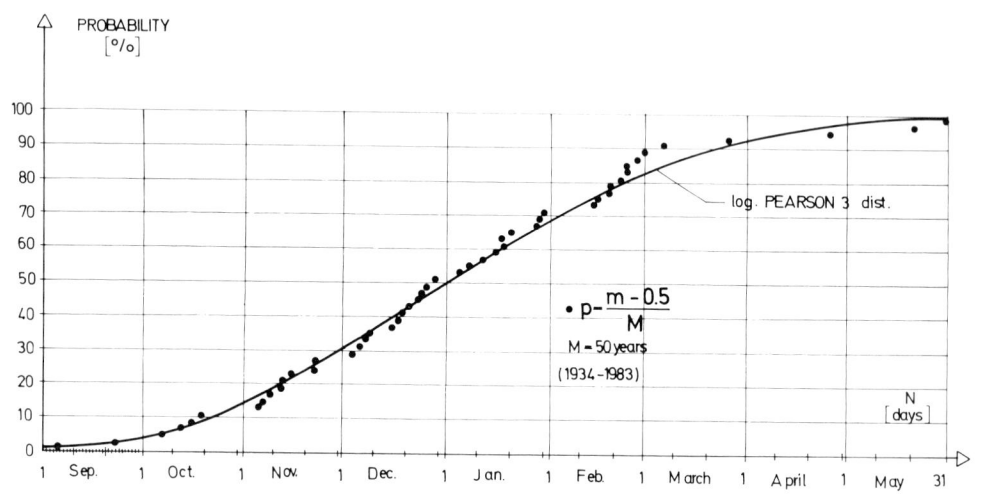

Fig. 7.1. Probability curve for the date of occurrence of the maximum flood level in the Konavosko Polje (Yugoslavia)

Hydrologic Budget for the Poljes in Karst 137

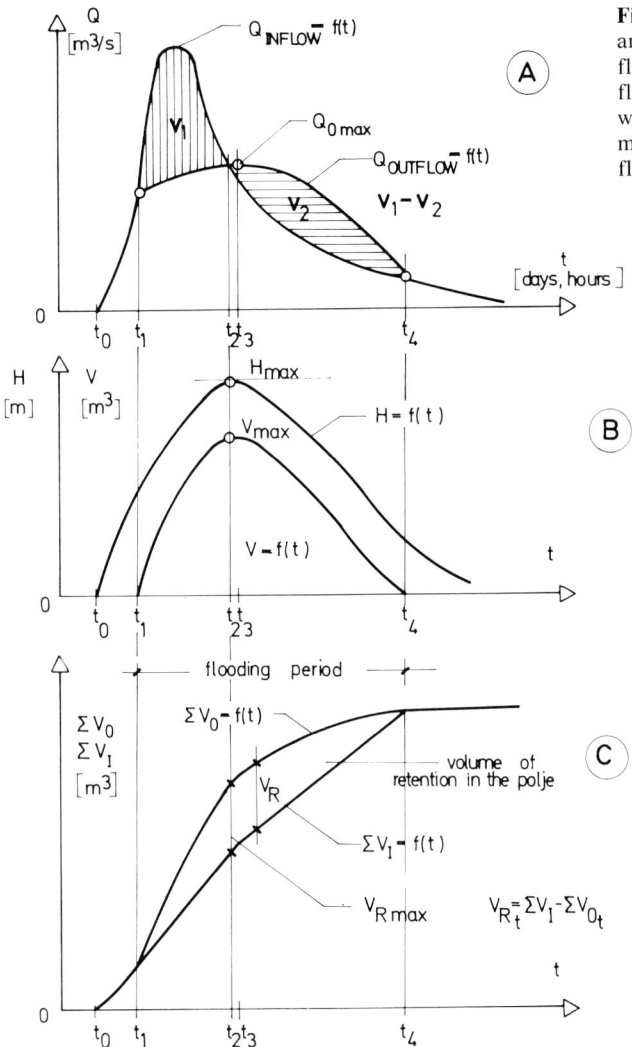

Fig. 7.2 A–C. Analysis of inflow and outflow hydrographs in the flooded polje. **A** Inflow and outflow discharge hydrographs; **B** water level hydrograph in polje; **C** mass curves of the inflow and outflow volumes

and the curve of the water volume in the polje retention $V = f(t)$, whereas Figure 7.2C gives the mass curves of the inflow and outflow volumes. The inflow capacity is equal to the outflow capacity until time t_1. After that moment the capacity of outlet structures and ponors is smaller than the inflow quantity of water and thus, causes flooding of the polje which lasts until t_4 when the flooding ends. From instant t_1 to t_2 inflows are greater than outflows and after instant t_2 the situation is inverse. The maximum retention volume occurs at time t_2. Floods in the polje can be observed by measuring the inflow quantities and the oscillation of the water level in the pre-ponor retention. The outflow quantities can be obtained by using the budget equation given the curves of the retention volume. It is not so simple in practice as it is difficult to control accurately all input data in a system. This situation becomes more complex when estavelle start acting like a spring. The ratio between the maximum inflow discharge $Q_{O_{max}}$ and the maximum inflow discharge $Q_{I_{max}}$ was defined by Ristić (1976) as the index of the retention effect f:

$$f = (Q_{O_{max}}/Q_{I_{max}}) < 1.0 \ . \tag{7.1}$$

That relationship is always smaller than *l* and in the poljes of Yugoslav karst it ranges from 0.1 to 0.6. The value of the index f indicates the flood level and duration.

The general form of the water budget equation for the poljes in karst is as follows:

$$\sum_{i=1}^{3} Q_{I_i} - \sum_{j=1}^{3} Q_{O_j} = \sum_{k=1}^{2} (V_k/\Delta t) \ . \tag{7.2}$$

Q_{I_1} denotes the inflows into the polje expressed in m³ s⁻¹ during the period Δt from the horizons not belonging to its own topographic catchment. These inflows are primarily from the higher horizons, but the inflow or even the flooding may be caused by a rise in a regional karst water table (Brook and Ford 1980). Such inflows are very difficult to control and measure. They represent the majority of all inflows, particularly in small, low lying poljes. Q_{I_2} represents the inflow from its own topographic catchment. Q_{I_3} is the inflow into the polje from the other catchments through hydrotechnical structures (tunnels, pipelines, pumping stations, etc.). These inflows can be easily measured and controlled. Q_{O_1} is the outflow from the polje by sinking through a system of small karst fissures and ponors of various dimensions. Section 5.2 deals with the procedures used to determine the outflow quantities through the ponors. Q_{O_2} represents the water losses by evapotranspiration from the topographic catchment area of the polje. Q_{O_3} is the outflow from the hydrotechnical structures (tunnels, pumping stations, etc.). ΔV_1 is the variation in the volume of the surface water in the polje expressed in m³, and refers to the period of time Δt. ΔV_2 represents the changes in the volume of the groundwater in the topographic catchment of the polje. While the changes in the volume of the surface water can be quite simply controlled, it is necessary to install a system of piezometers in order to measure the changes in the volume of the groundwater. In addition, it is necessary to know the change of effective porosity with depth. As the time period Δt becomes shorter, it is possible to define the budget more accurately. On the other hand, in that case, the hydrometeorologic measurements should be more frequent which makes the procedure more expensive and complicated.

Figure 7.3 shows the procedure for the determination of the maximum inflow discharge by measuring the variations in the water level in the polje (Fig. 7.3B), and by using the curve of the polje volume (Fig. 7.3A). The maximum value of the ratio $V/\Delta t$ should be defined, i.e. the maximum increase in the volume of the stored water in the polje ΔV. The total outflows from the polje occurring in that period should be added to the discharge defined in this way. In the case presented in Figure 7.3C, there were no losses, so that the value (quantity) of 72.9 m³ s⁻¹ can be taken as the total maximum inflow into the polje. The same procedure can be used to determine the maximum outflows from the flooded poljes in the karst.

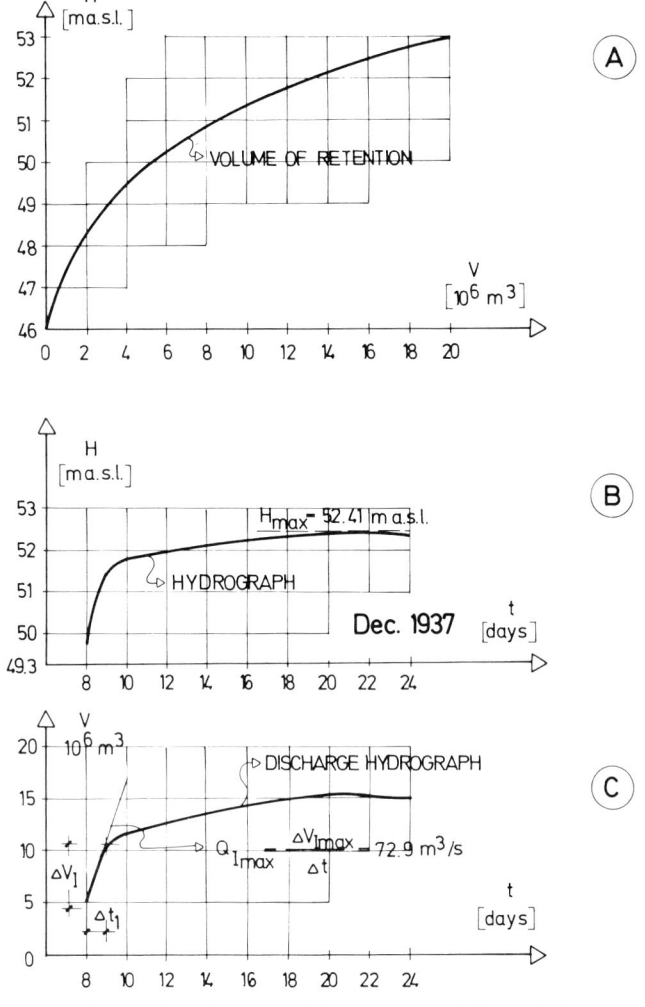

Fig. 7.3 A–C. Definition of the maximum inflow discharge at the example of the Konavosko Polje (Yugoslavia). **A** The curve of the polje volume; **B** water level in the polje; **C** discharge hydrograph

Yevjevich (1955) suggests four methods for the determination of the approximate water budget of the poljes in karst by measuring the inflow and outflow quantities. In some situations it is easier to control the inflows than the outflows, and, frequently, only partial control is possible. The budget should be worked out for the hydrologic year. Thus, the last term of Eq. (7.2) is eliminated since the computation is supposed to end and begin with the same minimum reserves of groundwater and surface water. In those poljes situated on cascades, it is possible to establish the relationship between the inflow quantities into the analyzed polje and the water quantities flowing out through the ponors located at higher horizons. Žibret and Šimunić (1976) suggest a simple, quick and approximate method for the determination of the budget in closed and flooded poljes. The method is practical since it requires standard hydrometeorologic data which are most frequently easily available. The starting assumption is that the total outflow from the polje Q_O depends upon the index of the antecedent precipitation IAP, the water level in the polje H and the period of the year M. The index of antecedent precipitations is determined by the equation:

$$\text{IAP} = \sum_{t=1}^{60} P_t e^{-\beta t}, \tag{7.3}$$

where P_t is the mean value of the daily precipitation in the catchment t taken before the computation on the outflow Q_O, whereas β represents the coefficient determined experimentally. The parameter of the season M is determined by the sequence of months. It defines the cyclical filling, i.e. emptying of the underground karst aquifer. The following is a multiple linear regressive equation for the Nevesinjsko Polje (Yugoslavia) whose catchment area is 485 km^2 and uses the theory of the least squares:

$$Q_O = 58.5 + 0.0279\, H - 5.89\, M - 0.258\, \text{IAP}, \tag{7.4}$$

with the coefficient of the multiple correlation $R = 0.87$. The budgets for other poljes yielded results of the same degree of accuracy. The applied model was used to explain 75 to 90 variations of the dependent variable Q_O. This result can be considered satisfactory, particularly since it does not call for any special hydrometeorologic measurements, and is simple in its application.

In the medium- and small-sized closed poljes the inflow quantities Q_I can be accurately controlled, the outflows from the polje Q_O can be determined using the following equation:

$$Q_O = Q_I - 0.0579\, A_2\, (H_3 - H_1), \tag{7.5}$$

where Q_O and Q_I are given in m^3 s^{-1} whereas the water levels H_1 are given in m and the flooded polje area A_1 is expressed in ha. Equation (7.4) refers exclusively to those cases in which the water level is measured once a day (always at the same time) and the indices refer to the first, second and third day of measurements. Equation (7.4) has been determined by applying Newton's approximation (up to the difference of the second order) for defining the curve which passes through three points. According to this equation, it is possible to define the unknown outflow from a given inflow or vice versa.

8 Water Temperature in Karst

8.1 Introduction

Measurements of water temperature in karst make it possible to answer numerous questions related to water circulation, its origin, retention time underground, etc. In order to determine thermal changes such measurements should be carried out with sensitive equipment, since these variations have very small amplitudes, i.e. very often, a few tenths of 1 °C. According to Komatina (1984), favourable conditions exist in karst for the application of geothermal investigations while detecting and monitoring the spatial position of the groundwater flows; therefore, both the karst mass and the groundwater flow cause temperature anomalies measurable even at the surface of the terrain. Komatina (1984) reported that the anomalies depend upon the following factors: (1) temperature of the groundwater; (2) depth, settlement and size of the groundwater flow; (3) degree of karstification of the medium between the flow and the terrain surface.

Undoubtedly, the main interest in hydrology and engineering practice related to water temperature includes the ecologic aspect of the problem, i.e. the possibility of using the water for drinking. This approach presents experience in using data on water temperature in karst for the solution of numerous theoretical and practical engineering problems. The emphasis is on the hydrologic aspect of the investigations and the use of temperature data for the explanation of hydrologic processes. It is evident that the same data can be used for numerous purposes.

8.2 Groundwater Temperature in Karst

Since heat is a conservative quantity in the subsurface environment, groundwater temperature can be used as a tracer to reveal the regional structure of a groundwater flow system. The groundwater temperature is an easily measurable element in a hydrologic survey and it is appropriate for a field study in an uninstrumented groundwater basin (Kayane et al. 1985).

The air temperature varies significantly during the year in the moderate belt form −30° to +50 °C. The soil temperature varies slightly depending upon its composition and depth and always has a time delay. The groundwater temperature in the karst zones situated inland varies very slightly. According to data from publications (Gunn 1983; Drogue 1980; Borić 1980; Kogovšek 1982; Sket and Velkovrh 1981; Komatina 1984; Cowel and Ford 1983; Karanjac and Altug 1980; Habič 1982; Sweeting 1973; Ede 1973; Petrik 1961), the water temperature in karst varies from a minimum of 4.9 °C to a maximum of 17.8 °C. Considering the limited data, the temperature range is somewhat greater, therefore, the temperatures may range from 4° to 20 °C.

The most influential factor affecting the water temperature in karst is whether a given groundwater mass belongs to the component of the concentrating mechanism or the input mechanism. Understandably, the water temperature is first affected by the soil temperature when the atmospheric water comes into contact with the soil. By the lateral water circulation through the soil, and particularly, by its flow through the subcutaneous zone, the water temperature takes on more and more of the numerous characteristics of the soil temperature. If the water transportation is effected by a fast turbulent flow through a developed system or the so-called conduit flow, the temperature variations of the water can be significant, since the effect of the medium is short, due to a short period of contact with the water. Water, sinking into a deep vadose zone and retained there for a time and then again flowing to the surface by a slow diffuse flow, has only slight temperature variations during the year. According to data published earlier, it can be stated with some certainty that these temperatures range from 8° to 16 °C.

Gunn (1983) used the data on water temperature in New Zealand karst to determine to which outflow component the water sample, taken at a given point, belongs: open streamflow, cave, jama, vertical fissure, aerated zone, saturated zone, subcutaneous zone, etc. Accordingly, he concluded that the air temperatures in the zone of origin have a strong influence on those groundwater flows which are fast and are retained underground for a short period of time. Then, the temperature of water flowing laterally in the layer immediately below the surface varies significantly, ranging from 4.9° to 14.7 °C, whereas the variations in the vadose flow are slight and range from 8.3° to 12.5 °C for the Mangapohue area, and only 12.5°–13.4 °C for the Glenfield area. The temperature of water flowing through the subcutaneous zone varies from 9.4° to 12.2 °C, and the temperatures of the flow sinking vertically along the sides of the jama vary from 9.3° to 12.8 °C.

The stated facts point out the possibility of using the data on water temperature in karst as an excellent indicator or the hydrologic processes, i.e. as important indicators in identifying the water circulation in the karst catchment.

Kogovšek (1982) studied the water flow through the vault of the Planinska Cave. The vault thickness was ca. 100–150 m. It approximately corresponds to the vertical flow of water. Whereas the air temperature in the open ranges from −15° to +25 °C, the air temperature in the Planinska Cave varies from +3° to +16 °C, and the temperature of the water sinking from the surface into the jama varies from only +6° to +12 °C.

Evidently, the thermal capacity of the karst mass has a great influence on the fast change and equilibrating of water temperature variations. It is important to note that these processes are of short duration, only 1 to 2 h, depending upon the soil humidity.

The specific features of the karst conditions were studied by Sket and Velkovrh (1981). They dealt with the temperature measurements of the groundwater karst river flowing through the system of the Planinska and Postojna Caves. Owing to the dimensions of these caves it is possible to reach the river banks and to carry

Groundwater Temperature in Karst

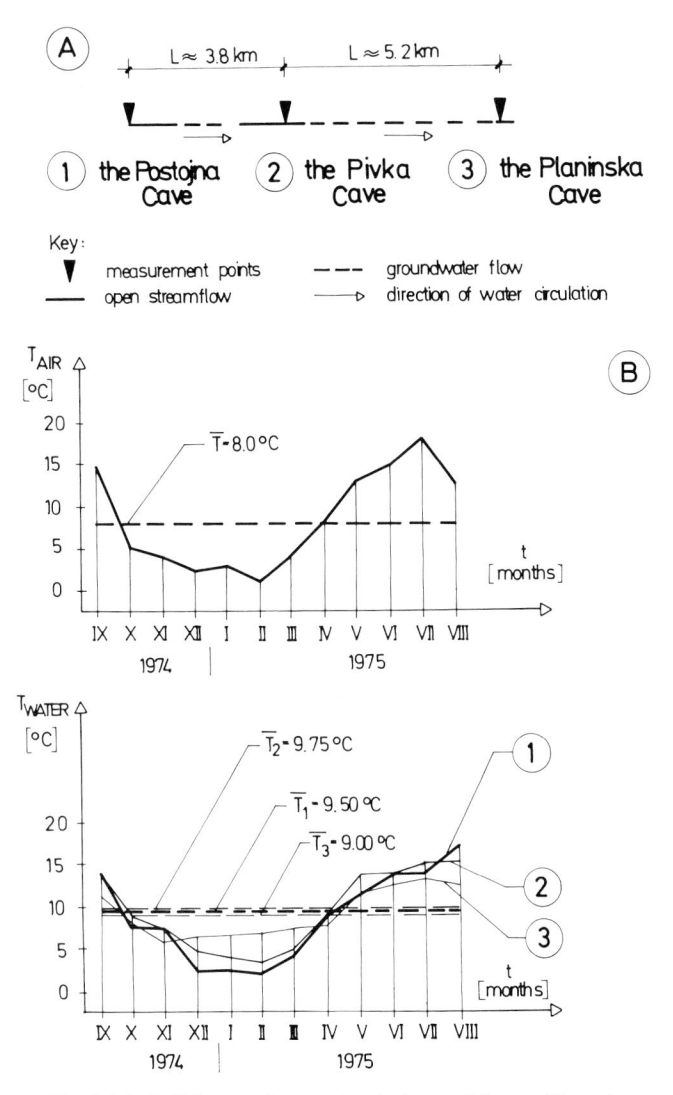

Fig. 8.1 A, B. Scheme of water circulation and the position of measurement points (**A**) and mean monthly air temperatures and water temperatures in the Postojna-Planinska cave system (**B**) in the hydrologic year 1974/75. (Sket and Velkovrh 1981)

out the measurements of air temperature in the caves and of the water temperature in the karst underground river. Great quantities of water heat the underground air in summer and cool it in winter, whereas when the discharge is smaller, this process is rather alleviated, and the annual water temperatures average about 9°C. Figure 8.1 shows the results obtained by measuring the mean monthly air temperature in the Postojna Cave and the water temperatures measured in three locations (according to Sket and Velkovrh 1981). Evidently, the longer retention of water underground leads to a more significant equilibration of the yearly temperatures. The mean monthly water temperatures in the first jama varied from 2° to 17 °C, in the second jama the variations were smaller, from 4° to 16 °C, and in the third, temperatures were the lowest, from 6° to 12 °C. These measurements prove that there is a strong influence of the karst mass on the water temperature changes when the turbulent flow has a velocity ranging from 0.1 m s^{-1} to 0.3 ms^{-1}. In this actual example the analysis of parallel thermograms obtained in the three jamas made it possible to determine the travelling time of water from one measurement point to another. The travelling time varied depending upon the water discharge, and became shorter with the increase in the discharge. The travelling time from the Pivka Cave to the Planinska Cave was 4 h for a 50 m^3 s^{-1} discharge, whereas for the discharge of 4 m^3 s^{-1} the travelling time was ca. 40 h.

The depth of the karst rock mass affected by the variations in the air temperature extends only a few meters below the surface of the soil. The zone of the thermic influence is predicted to deepen to several tens and even 100 m due to the fast infiltration and percolation of the surface water into a cracked karst medium. Drogue (1980) studied the thermic behaviour in two adjacent piezometers according to the scheme presented in Fig. 3.9 in karst in Terrien (France). Piezometer $\mathbb{T}_2$, intersects a karst channel through which the water from the surface sinks underground quickly and directly. In piezometer $\mathbb{T}_1$ there is no similar contact, which leads to essential differences in the thermic behaviour of water between the two piezometers. When the groundwater levels are low and when both locations have a laminar diffuse flow, there are practically no differences in the thermic profiles. The difference appears immediately following intensive rainfall when the water is significantly cooled and at the same time the groundwater levels are raised in the profile of the karst channel, i.e. in the zone of more rapid water circulation.

Borić (1980) studied the identical behaviour of the groundwater temperature in order to locate the zone with water losses from the Buško Blato Reservoirs, in the Dinaric karst in Yugoslavia. In the reservoir with the maximum volume of 800×10^6 m^3, with an average depth of 8 m and a maximum depth of 15 m, the water temperature ranges in summer up to +22 °C and in winter to +4 °C. As the storage is not very deep there is no change in the temperature considering the height of the water column. The storage was constructed in the karst zone by separating the ponor from the stored water by embankments and grouting curtains were built in areas of the largest cracks. Evidently, a satisfactory level of imperviousness was achieved, but it was noted that the water, nevertheless, was sinking underground through certain zones which were impossible to define using conventional procedures of geologic, hydrogeologic and hydrologic measurements. Therefore, the water temperatures were measured simultaneously in the basin and in several piezometers supposedly situated in the main routes of the water losses. These measurements were used to define exactly the zones through which the water sinks from the basin into the karst underground. The water temperature in the piezometer through which the water was sinking is presented in Figure 8.2. In the piezometer through which the water was sinking from the basin, the strong effect on the change in the groundwater temperature was noted. In summer the groundwater was essentially heated and in winter cooled. The groundwater temperature did not vary significantly and, during the entire year,

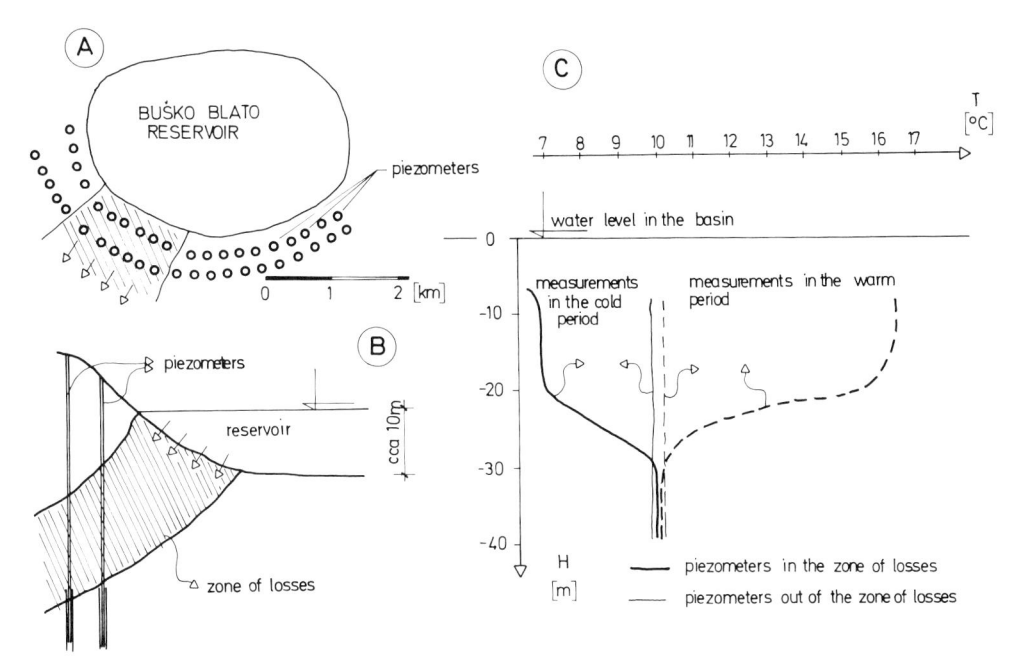

Fig. 8.2 A–C. Definition of the zone of losses from the Ručko Blato Reservoir (**A**) schematic situation (**B**) cross-section through the storage basin (**C**) diagram of the groundwater temperature in piezometers. (Borić 1980)

stabilized at ca. 10 °C in those places where there was no contact with the water from the basin. This method can be applied for the determination of the zones of loss, both in width and in depth. This task can be carried out only if the water temperature in the basin changes seasonally, if it essentially differs from the groundwater temperature which has to be approximately constant and if there are sufficient piezometers along the basin edge.

8.3 The Water Temperature of Springs and Open Streamflows in Karst

The karst spring is a natural path for the groundwater to the surface of the lithosphere through the hydrologically active fissures of the karst mass. Karst springs generally appear where there is contact between the carbonate rocks and flysch. The water comes to the surface through the porous rocks, which are practically insoluble, and sometimes not karstified (Bögli 1980). Pitty et al. (1979) stated three causes influencing the variability of the water temperature in the karst. First, the air temperature and the soil temperature during rainfall can lead to seasonal variations of the spring water temperature. These variations are particularly evident in shallow karst and in fast turbulent flows along privileged karst routes. Second, when the spring reaches the surface through a fissure system completely filled with water, the external air temperature can affect the water temperature through the caves or through the surface layer of soil or stones covering the

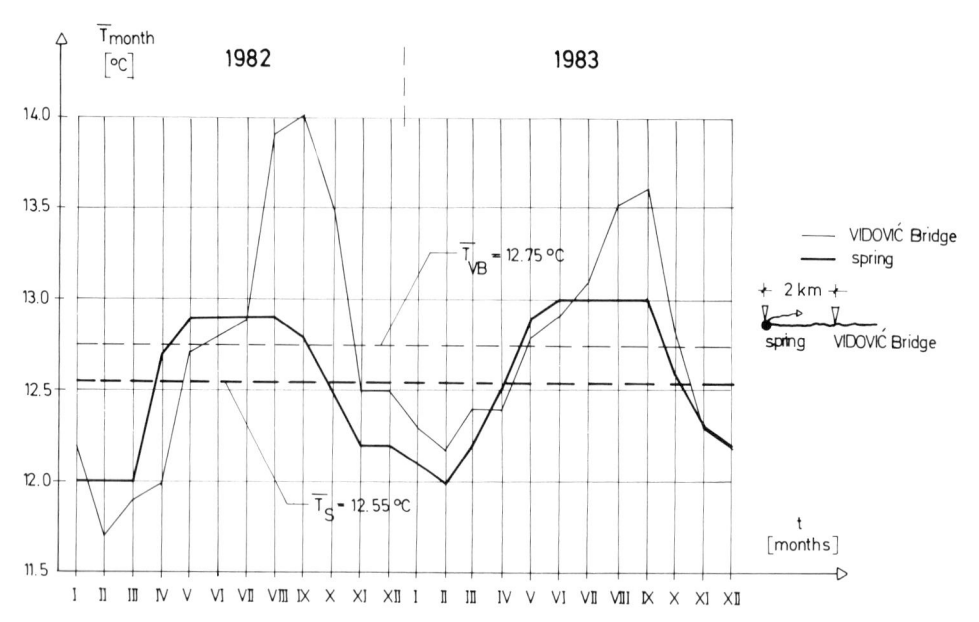

Fig. 8.3. Variations in mean monthly temperatures of the Jadro River at the spring and on the profile Vidović Bridge in 1982 and 1983

opening of the spring. The third possibility is the least frequent and has the weakest effect on the change of the spring water temperature. The influence of the outer temperature is effected through the voids filled by air lying above the main conduit of the spring. The change in temperature is caused by slow air circulation in the caves when there is contact between the cave air and the atmospheric air.

According to the previously mentioned facts, it is evident that there are seasonal variations in the water temperature in all of those springs which depend upon how direct the contact is between the groundwater and temperature of the atmosphere. When, however, the groundwater emerges via the spring, it is to be expected that its temperature will vary more under the influence of the air temperature variations. Figure 8.3 shows the variations in the mean monthly temperatures of the Jadro River on two profiles. At the spring itself the temperature variations are relatively slight, ranging from 11° to 14°C (for a longer time period). In the river canyon, the water flows with a minimum mean profile velocity of 1.5 m s^{-1}.

Table 8.1 presents the data of mean temperatures measured over several years.

Table 8.2 presents the same data for nine water-gauging stations on the rivers in the Dinaric karst for the same period. The differences in the temperature range, and their variations appear to be enormous.

Table 8.3 shows that the mean temperatures in the rivers measured over several years increase in proportion to the distance from the karst spring.

These considerations are completely in accordance with the results obtained in the investigations carried out by Pitty et al. (1979), shown in Figure 8.4.

The Water Temperature of Springs and Open Streamflows in Karst

Table 8.1. Characteristics of the spring temperatures in the Dalmatian karst measured in the period 1979/80 and 1984/85 (Yugoslavia)

Ordinal No.	Spring	T [°C]	T_{min} [°C]	T_{max} [°C]	σ [°C]
1	Ombla	12.45	10.33	15.33	1.98
2	Ljuta	12.06	9.57	14.07	1.75
3	Jadro	12.55	10.77	15.02	1.38
4	Žrnovnica	12.42	10.05	14.73	1.67
5	Vukovića Spring	10.00	8.62	13.12	1.32
6	Krka	10.38	8.12	12.52	1.52
7	Krupa	11.12	9.10	13.73	1.77

Table 8.2. Characteristics of temperature in the open streamflows in the Dinaric karst measured in the periods 1979/80 and 1984/85 (Yugoslavia)

Ordinal No.	River	Distance from the spring [km]	T [°C]	T_{min} [°C]	T_{max} [°C]	σ [°C]
1	Cetina	57	13.26	6.88	18.60	4.08
2	Zrmanja	42	12.73	9.40	16.48	2.86
3	Zrmanja	52	13.08	8.58	18.25	3.35
4	Krka	20	11.00	7.88	14.02	1.99
5	Krka	68	12.30	7.55	19.57	4.44
6	Neretva	152	12.18	6.96	17.06	3.20
7	Neretva	203	12.80	7.30	18.18	3.87
8	Neretva	212	13.46	7.34	18.34	3.80
Mean value			12.60	7.74	17.56	3.45

Table 8.3. The influence of the increased distance from the spring on the increase in the mean water temperature on the rivers in the Dinaric karst (Yugoslavia)

River	Distance from the spring [km]	Mean water temperature over several years [°C]
Zrmanja	9	9.00
	42	12.73
	52	13.08
Krka	0 (spring)	10.39
	20	11.00
	68	12.30
Cetina	8	8.40
	47	10.40
	57	13.26
Jadro	0 (spring)	12.55
	2	12.80
Neretva	152	12.18
	203	12.80
	212	13.46

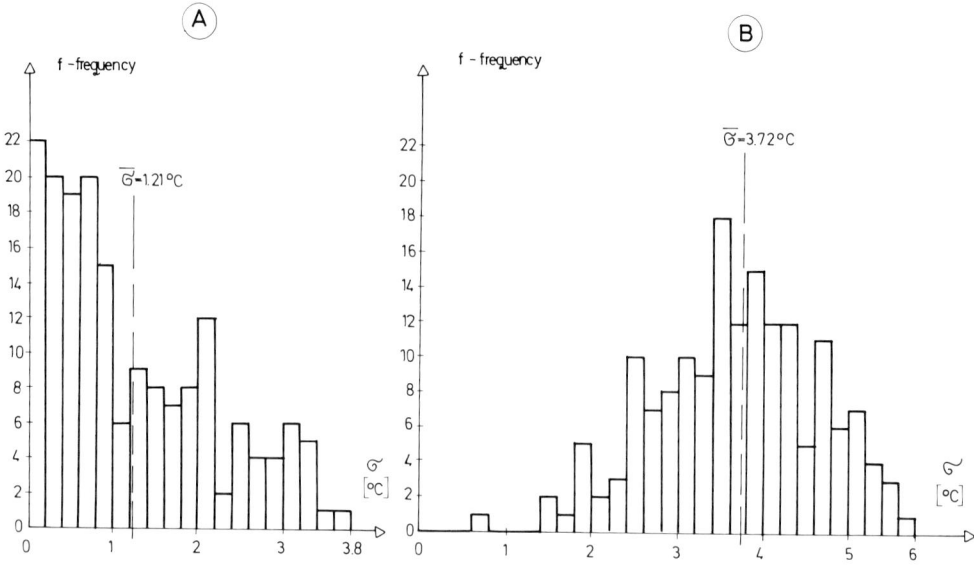

Fig. 8.4 A, B. Standard deviation of water temperature in the springs (**A**) and in open streamflows (**B**) measured in the Pennine Region (G. Britain). (Pitty et al. 1979)

Generally, the standard variation in the springs varies from 0° to 3 °C with a mean value of 1.5 °C, whereas the variations in the water temperature in the open streamflows in karst can be described by a symmetric distribution, and the standard deviations vary from 1° to 6 °C with the mean value of standard deviation of 3° to 4 °C.

The chronographic analysis of the water temperature of the karst springs can be used to determine whether the springs are conduit springs, i.e. the water flows from the underground to the surface by fast and turbulent flow through a developed system of fissures or whether they are diffuse springs, i.e. the water reaches the surface by gradual and slow percolation through small karst cracks. The same results have been obtained and identical conclusions drawn by Sweeting (1973); Ede (1973); Novak (1971) and Cowell and Ford (1983). They have all found that on conduit springs, the water temperature and hardness vary a great deal and correlate highly with the air temperature, whereas in the diffuse springs the temperature variations are slight during the year, only 2° to 3 °C, and their correlation with the air temperature is not significant. These conclusions can be applied to all springs regardless of whether they are located in areas with cold, moderate or warm climates.

Cowell and Ford (1983) found that in the Ontario area (Canada), the standard variation of the water temperature is 5.9 °C for conduit springs, and 1.32 °C for diffuse springs. Ede (1973) found that the water temperature measured during one year in conduit springs in South Wales ranged from 6.5° to 11 °C, and in diffuse springs from 10.2° to 10.9 °C and from 11.0° to 12.1 °C. Novak (1971) reported the water temperature variations in the diffuse springs in the Slovenian karst to be from 7.5° to 9 °C, and in the conduit springs these temperature ranged from 3.5° to 14.2 °C.

Precise, continuous measurements of the water temperature of the karst springs, with a simultaneous monitoring of the spring hydrographs, make it possible to define the volume of the large fissures directly connected to the spring. Atkinson (1977) and Williams (1983) suggested a method which yields identical results, but included the observation of the changes in the water hardness. The water retained for a longer period of time underground has a constant temperature and hardness. Immediately following abundant rainfall the formerly retained water flows out through the system of cracks. That water has a different hardness and temperature from that of the rainfall water. The volume of the system of fissures can be defined by the function of the groundwater level before the rainfall, if the temperature and hardness are accurately measured; or, at least one of them, with the integration of the hydrograph of the spring in the period beginning with its increase to the beginning of the change in temperature. Summer rainfall brings warm water, and winter brings colder water into the underground karst system.

Karanjac and Altug (1980) measured and analyzed the water temperatures of about 25 medium-size karst springs downstream of the Oymapinar Dam (Turkey). These data were evaluated to indicate the depth of circulation and the extent of karstification. The temperature fluctuations (minimum, maximum and amplitudes) were used to formulate the hypotheses related both to the recharge sources and to the depths and lengths of the groundwater circulation. The main goal of these analyses was to evaluate the leakage of the Oymapinar Reservoir. Karanjac and Altug (1980) suggested that small amplitudes of temperature recorded at this group of springs during the whole year indicated deep sinking of the groundwater (more than 100 m below the surface) and its long path through the underground. If the temperature at a group of springs changes quickly and directly (most frequently immediately after heavy rains), this means that the underground circulation takes place near the surface (not deeper than 10 to 50 m), and that its path, starting with the infiltration of water until its appearance at the karst spring, is short and direct. These data can be very significant in the design of grouting curtains on the reservoirs in the karst.

Submarine springs represent a specific case of karst springs. Their capacities can be at least partly determined by measuring the water temperatures in them and in the surrounding sea. Alfirević (1969) and Petrik (1961) carried out numerous measurements of temperatures in the submarine springs of the Adriatic Sea. It has been shown that the water temperature on the very location of the submarine spring varies slightly from the surface of the sea to the bottom of the submarine spring. The submarine springs generally exhibit vertical homothermy, whereas the sea never or very rarely has homothermic conditions. The difference between the water temperature of the submarine spring and the surrounding sea increases as the discharge from the submarine spring becomes greater. In the dry periods, when the capacity of the submarine spring essentially decreases or they completely disappear, the difference between the water temperature of the submarine spring and the sea can hardly be noticed.

9 Man's Influence on the Water Regime in the Karst Terrains

Living conditions are not favourable for man in karst regions. The main reason for this is the unfavourable water regime. In autumn and winter excessive precipitation often results in the flooding of fertile poljes and river valleys, which frequently last a long time. The water is retained in these fertile areas, more than 3 months on the average; the maximum retention lasting 6 months. Immediately after the floods end, long dry periods begin. Both phenomena adversely affect the agricultural production of the area. Since the soil is fertile and temperatures exceptionally favourable (particularly in the Mediterranean regions), the agriculture could be significantly improved by the regulation of the water regime. The previously mentioned facts have since ancient times forced man, living in these areas, to build more or less significant hydrotechnical structures. All of these structures were built with the same objective, i.e. to improve the water regime, and hence, the living conditions in the region. Today, intensive works are being carried out related to the regulation of the water regime in karst. Past experience has shown that many of these works have been suboptimal. The benefit resulting from these works in one area was frequently smaller than the damage caused in another area. This frequently occurred when the system was not thoroughly studied from hydrologic and hydrogeologic standpoints. Damage occurred most frequently in the lower karst horizons, but negative effects were also seen in the higher terrains. This damage was primarily caused by the water regime and floods. This book does not deal with the environmental impact of these effects. It can be stated with certainty that the environmental effects exerted by these structures in karst are even more serious, dangerous and threatening for man's existence than the consequences related to the technical and hydrologic aspects. Therefore, detailed interdisciplinary investigations must be effected when designing structures in karst. At the same time, it is necessary to predict what changes in the water regime the designed structure will exert considering a wider area. Finally, it is necessary to establish an extensive system for measuring all hydrologic and hydrogeologic phenomena which will serve to precisely assess the influence exerted by the structures in the water regime. It should be remembered that the groundwater levels affect all processes of water circulation in karst. This fact has been emphasized since the identification of the system has been frequently realized without a proper knowledge of the changes in the groundwater levels existing before and after the construction of the system. This applies particularly to the earlier structures built before 1960. Practically all available methods and systems can be used, without exceptions, for the detection of the system. All facts are pertinent including the economic aspect of the problem. An interdisciplinary approach to the investigations, as well as cooperation among experts engaged in various disciplines, are essential since it is necessary for successfully establishing the complex systems of water circulation in karst. According to our observations, two methods have been usually used when determining the effect exerted by the structure on the outflow processes: parametric-genetic or statistical methods or their combination. The statistical methods for defining the discrimination level can be used independently without the parametric-genetic methods, whereas the latter can hardly be used independently. The methodology applied to this field has not been sufficiently developed. Special emphasis should be placed on the application of discrimination analysis and factor analysis as well as some other classification methods which have not yet been used in hydrology. The hydrotechnical works carried out in karst are numerous, but can be categorized as follows:

1. Water storage;
2. Increase in the capacity of the outlet structures;
3. Surface hydrotechnical works;
4. Action of the groundwater;
5. Use of the karst springs water;
6. The development of the brackish karst springs.

9.1 Water Storage

Water storage in karst is primarily concerned with the construction of the surface storage reservoirs. In order to ensure their watertightness it has been necessary in most cases, to complete extensive grouting works. The construction of grouting curtains ensures the watertightness of the surface reservoirs, and at the same time it reduces or completely eliminates the communication between the groundwater in karst; thus, groundwater storages are formed simultaneously. The current trend is to form exclusively the groundwater storages (Paloc and Mijatović 1984). From the hydrologic aspect their influence on the environment is very similar or even identical.

There are a great number of surface reservoirs in karst. There are about 30 reservoirs with a volume of more than $30 \times 10^6 \, m^3$ in Yugoslavia. The largest reservoir is Bileća on the Trebišnjica River with a capacity of $1100 \times 10^6 \, m^3$. Large storage basins are generally built for the exploitation of water for the production of electric energy. They are multipurpose structures since they simultaneously protect the downstream areas from floods, while smaller quantities of water can be used for other purposes (water supply, irrigation, etc.). Frequently, the surface reservoir is situated 500 m or more above the hydroelectric power plant, thus ensuring a high quality of electric energy. Zötl (1984) described the construction of the Diessback Reservoir (Austria). The reservoir is at an altitude of 1400 m above sea level, i.e. 718 m above the Saalach Valley where the power plant is situated.

Mikulec and Trumić (1976) classified the storage basins in karst in the following way; (1) reservoirs situated entirely on the impermeable layers; (2) basins situated in partly or completely permeable karst rock mass; (3) reservoirs built in the karst poljes with impermeable or slightly permeable bottoms. The sides from which the water flows into the basin are impermeable, and in the outflow zone they are permeable; (4) reservoirs built in poljes with impermeable or partly permeable bottoms. The sides are protected by impermeable embankments or lined with impermeable materials.

The reservoirs belonging to the first group are generally small and are rarely found in karst. The watertightness of the reservoirs belonging to the second type is provided by a favourable geologic distribution of the underground impermeable layers (barriers). Water losses through the basin bottom and sides are frequent without regard to the natural conditions and the works carried out to improve the impermeability. The objective of the engineering works is to reduce these losses and, consequently, to make it possible to exploit the water from the basin economically and efficiently. The construction of a reservoir in karst affects the natural conditions of the water circulation. Consequently, its effects are seen particularly in the downstream horizons. The discharges of the springs situated downstream are either decreased or levelled. The minimum discharges are frequently increased and the maximum discharges are decreased, whereas the mean annual discharges remain the same. On the upper horizons, the water reservoirs in karst can revitalize the old emptying systems which have not been functioning because of the descendent of the karstification base. A few examples of the Yugoslav karst illustrate the consequences resulting from the construction of the reservoir as seen on the downstream and upstream horizons.

The Cerkničko Polje is situated in the catchment of the Ljubljanica River in Slovenia (Yugoslavia) in areas of karst covered by vegetation. The bottom of the polje covers an area of ca. 35 km^2. The greatest flood in the polje reaches an elevation of 552 m a.s.l. and floods an area covering 26 km^2. The water volume for the maximum flood is 70×10^6 m^3 (Gospodarič and Habič 1978). The average duration of the floods in ca. 240 days a year, and frequently the water is retained during the entire year in the lowest part of the polje. Therefore, it was decided to form a permanent reservoir in one part of the Cerkničko Polje to ensure the water supply, to construct a small hydroelectric power plant and to encourage the development of tourism, fishing and other commercial activities (Breznik 1983). The work on the reservoir began in 1969. The main ponors in the polje were closed. This action made the duration of the floods in the polje last about 75 days a year longer, but the objective, i.e. the formation of a permanent reservoir, was not achieved (Fig. 9.1). Figure 9.1 also shows that the maximum level was raised 1.5 m, and the lowering of the water level in the polje was decreased from 7 cm day^{-1} to 4 cm day^{-1}. The example of the Cerkničko Polje clearly shows that apart from the main ponors in the karst areas, there are numerous secondary and often invisible ponors which are very difficult to eliminate. Therefore, the work was continued by building dams and grouting curtains intended to isolate the ponor zones from the water of the new reservoir. The discharge of all the downstream karst streamflows, recharged by the water from the Cerkničko Polje, was reduced and some of these streamflows now dry up in summer. The final construction of the reservoir will reduce the mean annual discharge of the spring by about 70% (Breznik 1983). The Bileća Reservoir on the Trebišnjica River was built in 1967. After construction of the Grančarevo Dam, 123 m high, the largest artificial lake in Yugoslav karst with a volume of 1100×10^6 m^3 was formed. The

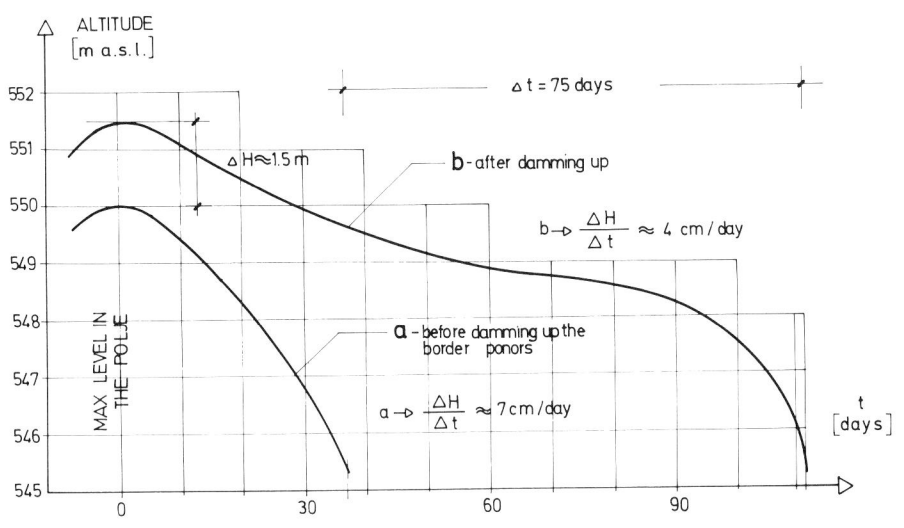

Fig. 9.1. Average recession parts of hydrographs of the Cerknica Polje (Yugoslavia); *a* under natural conditions; *b* after damming up the main ponors. (Gospodarič and Habič 1978)

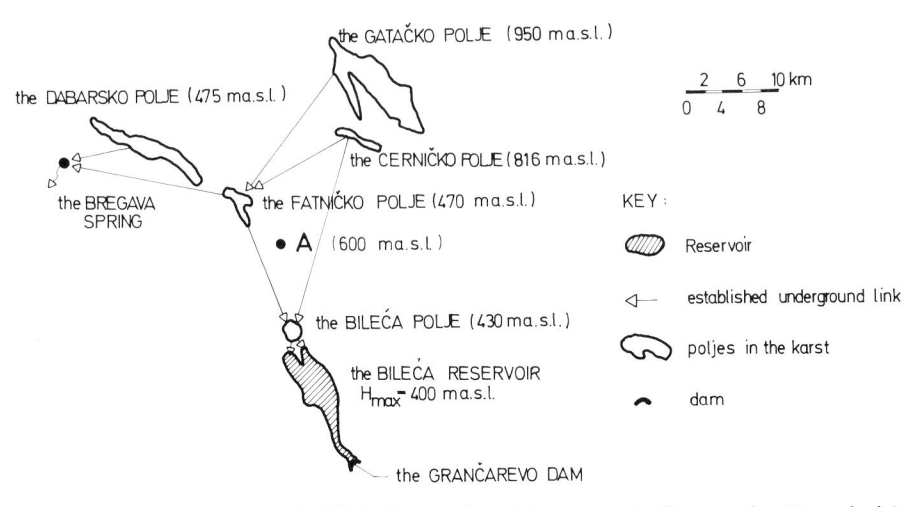

Fig. 9.2. Schematic map of the Bileća Reservoir and karst area in Herzegovina (Yugoslavia)

dam, the reservoir and the Trebišnjica River are situated in the region of bare Dinaric karst (Yugoslavia). Figure 9.2 presents a schematic situation of the reservoir and the surrounding poljes. The catchment area of the reservoir is difficult to determine, but it covers approximately 4000 km². The construction of this large artificial lake significantly affected the regime of the entire catchment, particularly the poljes situated at higher elevations. In the Bilećko Polje, with an average height of 10–20 m above the highest reservoir level, springs appeared along the entire polje in those places where they had not been previously recorded. Those springs occur when the high level of the water in the reservoir coincides

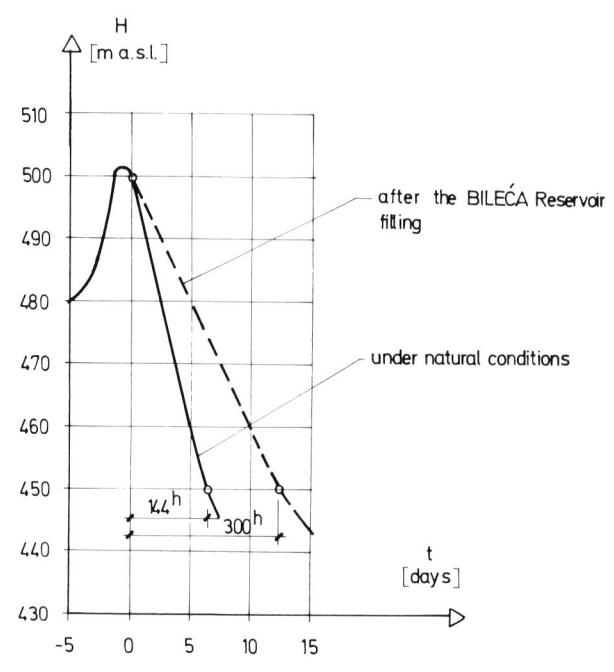

Fig. 9.3. Groundwater recession curve recorded on piezometer A in Fig. 9.2. (Milanović 1986)

with the occurrence of heavy precipitation in the catchment. The change in the natural levels of groundwater, caused by the reservoir's construction, makes it possible to reactivate the conduit system in karst which stopped functioning after the lowering of the erosion basis. Its reactivation creates numerous problems and the economic damage can be considerable. Figure 9.3 presents the groundwater recession curve recorded at piezometer A (Fig. 9.2) located 10 km from the edge of the Bileća Reservoir. Milanović (1986) found that the submergence of the spring zone of the Trebišnjica River affects the dynamics of the emptying of the karst aquifer. The high water levels of the Bileća Reservoir decrease the groundwater piezometric gradient. Consequently, the flow velocities are smaller in relation to those under natural conditions.

The Bileća Reservoir has affected the flooding regime in poljes 5 – 30 km distant, situated on horizons about 70 to 80 m higher, i.e. the Fatničko and Dabarsko Poljes. Table 9.1 presents the average number of days of flood duration per month and per year, as recorded in the period prior to the construction of the reservoir (1949 – 1967) and following its construction (1968 – 1982). Figures 9.4 and 9.5 give a chronologic review of the flood occurrences for different years. The duration of the floods was longer in both poljes.

Testing by the t-test and the Wilcoxon non-parameter test shows that there is a statistically significant difference between the lengths (time) of the flood duration. The maximum annual flood levels decreased in the Dabarsko Polje by 1.4 m, and in the Fatničko Polje by 1.8 m, but the flood duration in the poljes was longer. At first sight these results seem to be contradictory. They can be explained by the

Water Storage　　　　　　　　　　　　　　　　　　　　　　　　　　　　　　　　　　155

Table 9.1. The number of days of flood duration before and after the construction of the Bileća Reservoir in the Dabarsko and Fatničko Poljes

Period	I	II	III	IV	V	VI	VII	VIII	IX	X	XI	XII	Year
Dabarsko Polje													
1947–1967	19.2	14.2	15.7	10.1	6.4	0.6	0	0	0	4.8	19.5	20.3	110.8
1968–1982	19.1	17.0	17.9	17.7	12.3	4.3	0	0	2.7	7.7	14.9	22.1	135.7
Fatničko Polje													
1949–1967	20.2	14.7	17.8	16.7	9.5	1.4	0	0	0.4	5.9	20.8	22.9	130.3
1968–1982	20.9	18.1	18.8	21.5	12.3	6.1	0.5	0.2	2.7	8.9	15.4	23.1	148.5

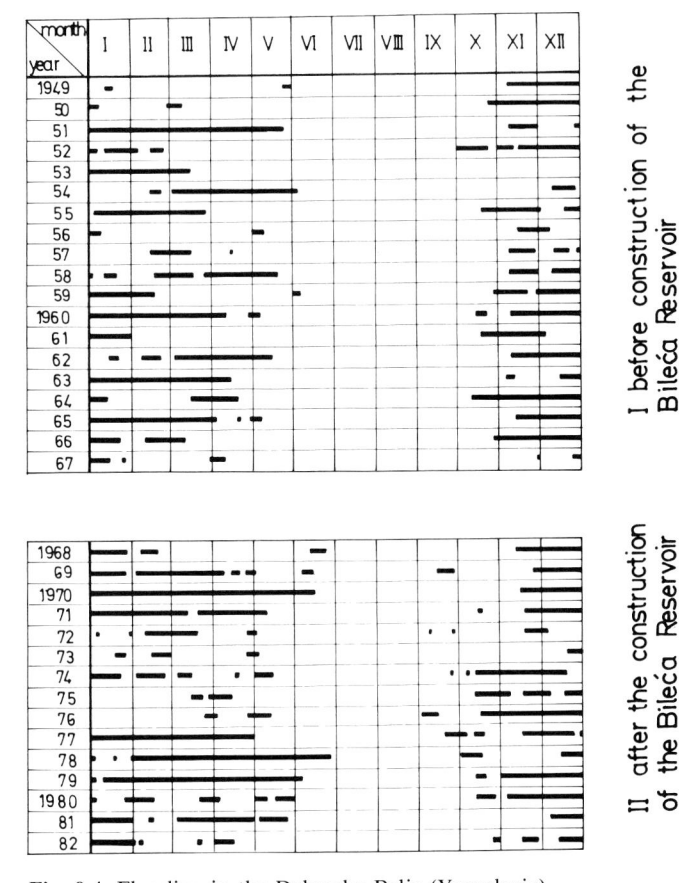

Fig. 9.4. Flooding in the Dabarsko Polje (Yugoslavia)

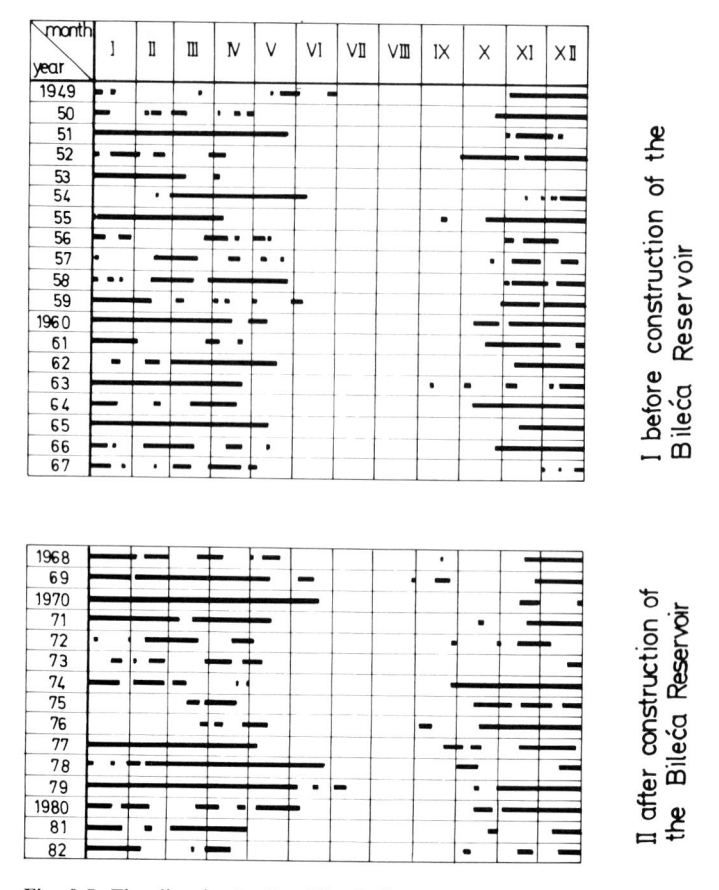

Fig. 9.5. Flooding in the Fatničko Polje (Yugoslavia)

regulatory effect of the reservoir on the water regime and by the general change in the ground levels in the entire catchment. It should be noted, however, that floods occurring in November under natural conditions lasted longer than after reservoir construction. November happens to be the month with the most intensive precipitation. Then, the underground karst pores are completely filled by water. The construction of the Bileća Reservoir and the raising of the groundwater levels opened new paths for water circulation in karst and reactivated fossilized paths. Consequently, when the underground cracks are being filled, most frequently in November, the water is retained in the poljes for a shorter time. Greater amplitudes of the oscillations in the groundwater levels activated the new cracks in the karst mass and intensified its effective porosity. Thus, the volume of the underground retention increased and the circulation of the groundwater was speeded up under conditions of low piezometric levels and low water levels in the Bileća Reservoir. Therefore, the maximum flood levels were decreased. The duration of the floods was longer because of the general raising of the groundwater level in the catchment.

Water Storage 157

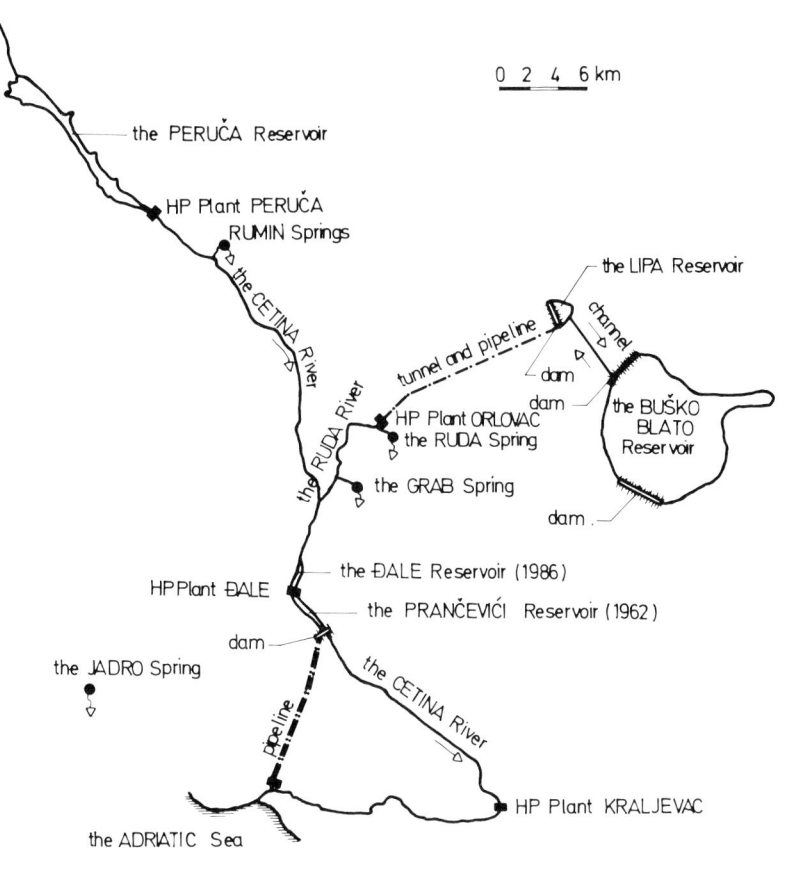

Fig. 9.6. Map of the Cetina River and reservoirs in its catchment area (Yugoslavia)

Miličević (1976) studied the influence of the reservoir on the change in the regime of the naturally flooded upstream poljes. He used the model of the multiple linear correlation link between the dependent variable, i.e. the maximum flood volume in the polje, and a series of independent variables. He used the hydrometeorologic parameters of the catchment as independent variables; the air temperature in the period of the flood increase, the water level on the upstream polje, the index of antecedent precipitation, etc. The equation employed for the computation of the independent variable under the new conditions has been defined by the theory of the least squares for that period before construction of the reservoir. The influence of the reservoir on the upstream poljes was established according to the agreement between the results obtained by measurements and computation. A system of reservoirs has been built in the catchment of the Cetina River (Fig. 9.6). Three reservoirs were built by damming the river. The largest reservoir, Buško Blato, was built in the Livanjsko Polje. Until the reservoir construction the waters from that polje were connected through underground paths with numerous springs on the left bank of the Cetina River. The largest springs are shown in

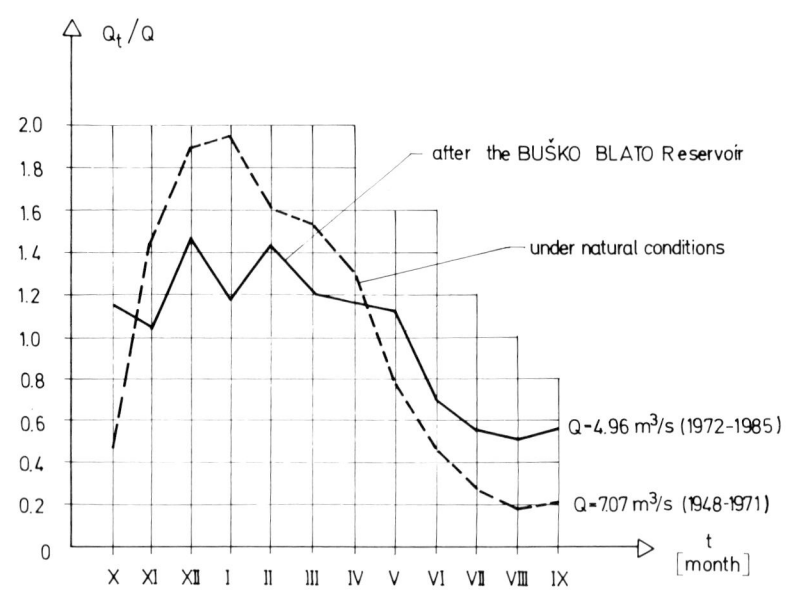

Fig. 9.7. Influence of the Buško Blato reservoir on the water regime of the Grab Spring (Yugoslavia)

Figure 9.6. The construction of this large reservoir (Buško Blato) with a capacity of 800×10^6 m^3 significantly affected the water regime of the springs on the Cetina River left bank. The mean annual discharge of the Grab Spring was reduced from 7.37 m^3 s^{-1} to 4.96 m^3 s^{-1}, i.e. 2.11 m^3 s^{-1} or 30% of the previous discharge. In addition, the water regime was levelled off as displayed in Figure 9.7. The modulus coefficients under the new conditions ranged from 0.5 to 1.4, whereas under natural conditions they ranged from 0.2 to 2.0. The minimum annual discharges of the Grab Spring were increased from 0.9 m^3 s^{-1} to 1.5 m^3 s^{-1}, whereas the maximum discharges remained unchanged. The Buško Blato Reservoir exerted a significant influence on the analyzed spring in karst. It has increased its minimum annual discharges. The spring discharge under the new conditions does not fall below 1 m^3 s^{-1}, whereas under natural conditions the spring often dried up. At the Rumin Spring the mean discharges over several years were decreased from 2.83 m^3 s^{-1} (1950–1971) to 2.21 m^3 s^{-1} (1972–1985), i.e. 0.62 m^3 s^{-1} or 22% of the discharge under natural conditions. The minimum discharges remained the same, whereas the maximum annual discharges considerably decreased from the average over several years of 12 m^3 s^{-1} (1972–1985) to 9 m^3 s^{-1} (1972–1975). The construction of the Buško Blato Reservoir levelled off the water regime of the Cetina River, which favourably influenced the degree of hydroenergy exploitation in all HP plants and particularly in those situated downstream.

The Jadro River Spring is 15 km air distance from the Prančevići Reservoir and about 40 km from the Buško Blato Reservoir (Fig. 9.6). The Prančevići Reservoir has a small capacity of 6.8×10^6 m^3, and was built in 1962. Even though it

Table 9.2. Mean typical discharges of the Jadro Spring from 1950–1985

	Period	Q [m³ s⁻¹]		
		Minimum	Mean	Maximum
I	1950–1961	2.2	5.4	45
II	1962–1971	2.4	6.3	48
III	1972–1984	2.9	8.8	75

Table 9.3. The increase in the inflow of the Jadro Spring ΔQ due to the construction of the Prančevići and Buško Blato Reservoirs

ΔQ	Minimum		Mean		Maximum	
	[m³ s⁻¹]	[%]	[m³ s⁻¹]	[%]	[m³ s⁻¹]	[%]
ΔQ_{I-II}	0.2	9	0.9	17	3	7
ΔQ_{II-III}	0.5	23	2.5	46	27	60
ΔQ_{I-III}	0.7	32	3.4	63	30	67

has small dimensions, it has influenced the change in the water regime of the Jadro Spring. Table 9.2 presents the mean minimum, average and mean maximum of several years of discharges measured over three periods. The first period covers the years between 1950–1961 and refers to the natural regime. The second period from 1962–1971 refers to the time when the Prančevići Reservoir was in operation. The third period, 1972–1985, includes the influence exerted by the Buško Blato Reservoir on the hydrologic regime of the Jadro Spring.

Table 9.3 shows that the minimum of several years of discharges increased 9% due to the construction of the Prančevići Reservoir and then another 23% because of the Buško Blato Reservoir construction. A greater increase was recorded in the mean of several years of discharges. Even the maximum average of several years of discharges have been increased. The results of all differences ΔQ as well as the percentages of the increase with respect to the discharge under natural conditions are given in Table 9.3.

The given examples (Table 9.3) show that the construction of a reservoir in karst can affect the hydrologic regime of the spring located further downstream in a number of different ways, and that accurate predictions and reliable conclusions cannot be drawn without numerous extensive and accurate measurements.

9.2 Increase in the Capacity of the Outlet Structures

The increase in the capacity of the outlet structures reduces the duration of the floods in general, primarily those occurring in the poljes in karst. The economic effects are achieved by the construction of a tunnel or some other structures to be used for the fast evacuation of water (spillways, excavations, pumping stations). The stability of the agricultural production depends primarily upon the

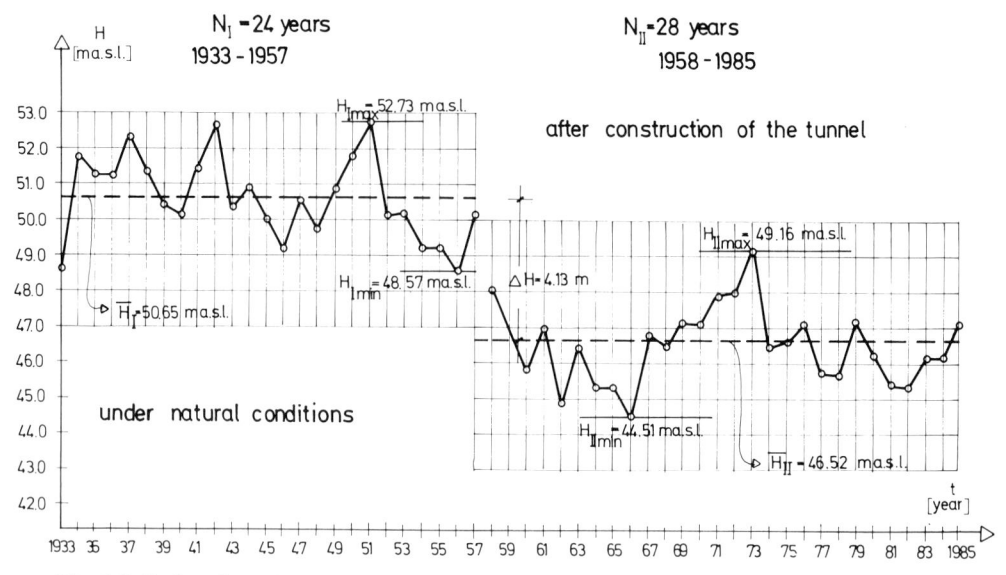

Fig. 9.8. Series of maximum annual water levels in the Konavosko Polje (Yugoslavia)

complete protection of the fertile region from floods. Therefore, since ancient times, man has attempted to increase the swallow capacity by widening and cleaning the ponor openings. Such attempts have usually been condemned to failure because the capacities of the ponors depend upon the size of the inflow only to a small extent.

In 1958 a tunnel delivering the water directly from the polje into the Adriatic Sea was built in the Konavosko Polje (Yugoslavia). The cross-sectional area of the tunnel is 13 km^2 with a maximum capacity of 40 m^3 s^{-1}. The construction of this tunnel significantly reduced the duration of floods and their elevation. Figure 9.8 presents the series of the maximum elevations of the annual floods in the polje both under natural conditions and after tunnel construction. The mean flood elevation was decreased for 4.13 m. Thus, the proposed objective of complete elimination of floods has not been achieved. The lowest and the most fertile part of the polje is still being flooded for a short period of time. In order to achieve the final objective, it is necessary to increase the capacity of the tunnel and to work out a detailed drainage network.

One of the most interesting poljes in the Dinaric karst in Yugoslavia is the Vrgorac Polje. The first tunnel for water drainage from the polje was drilled in 1938, and in 1974, it was widened as the floods were not eliminated. Their duration was reduced, but it was not sufficient for the demands of the intensive agricultural production of Mediterranean plants since the polje is located 25 m above the Adriatic Sea level. Figure 9.9 presents the map of the Vrgorac Polje with the position of the tunnel for the drainage of flood water. Until the construction of the tunnel in 1938, the floods were of long duration and reached maximum levels of 32.7 m above sea level. Since measurements were not taken before 1926, it was supposed, according to eyewitness accounts, that the floods exceeded

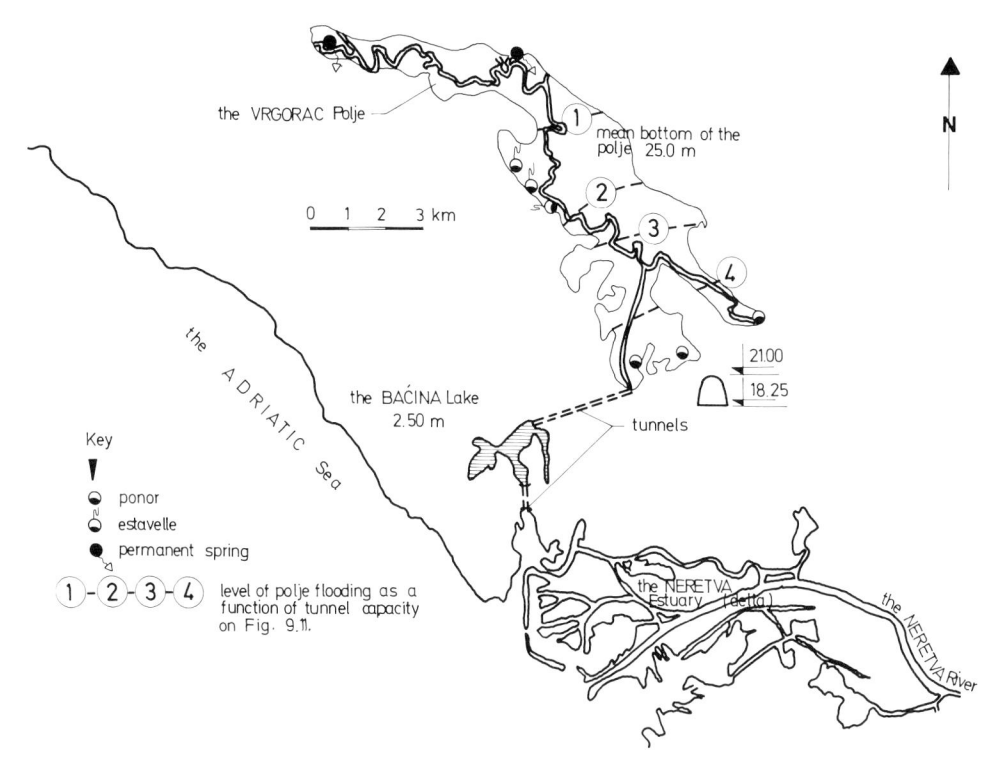

Fig. 9.9. Map of the Vrgorac Polje and tunnel with the position of flood levels at the different tunnel capacities (Yugoslavia)

33.5 m, which meant that practically the entire polje was flooded. Figure 9.10 shows a series of maximum water levels in the polje before and after tunnel construction. Until 1938, the average (observed) maximum level was 29.63 m above sea level, whereas after tunnel construction it was decreased to 3.09 m. Tests of differences of arithmetic means between the series of maximum annual levels in the polje performed by several parameter and non-parameter tests proved that the arithmetic means differ significantly. Thus, the influence of the tunnel on the floods has been statistically verified. The tunnel capacity was increased in 1974, and consequently, the maximum flood level was decreased by more than 2.31 m. The tunnel significantly reduced the flood durations, but did not eliminate them entirely since the tunnel's maximum capacity ranges to about 36 m^3 s^{-1}. Maximum average inflow discharges are the same during the three periods and vary about 95 m^3 s^{-1}, whereas the greatest maximum level reached 154 m^3 s^{-1}, as observed on 22 February 1969. This discharge is considered as a discharge of 100-year-return period. Understandably, it would not be economically justified to drill a tunnel with a 154 m^3 s^{-1} capacity, even if it would mean eliminating the floods, since the capacity of the open main streamflow of the Matica Vrgorska River does not exceed 50 m^3 s^{-1}. Figure 9.11 B shows that the transformation of the flood hydrograph (20 September 1966 – 5 January 1967) in the Vrgorac Polje

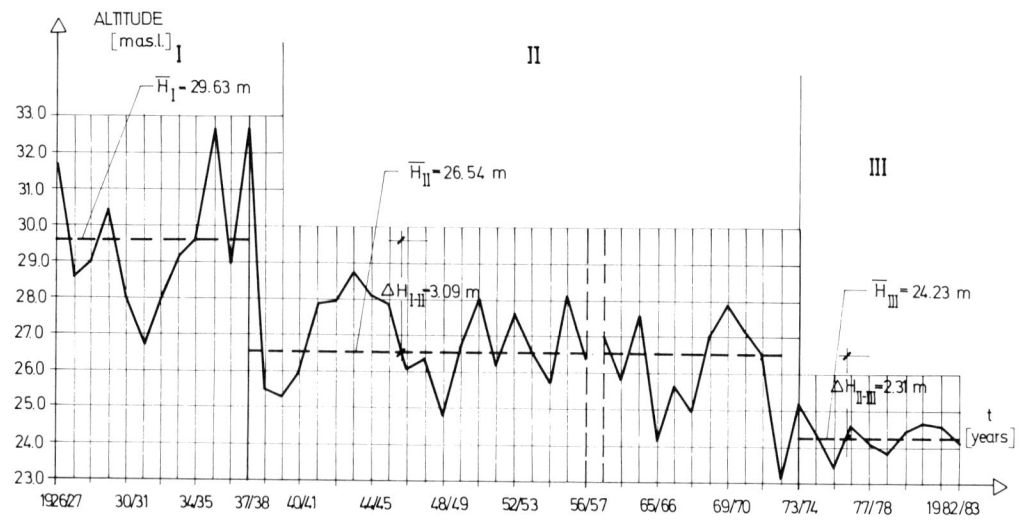

Fig. 9.10. Series of maximum annual water levels in the Vrgorac Polje (Yugoslavia). *I* before tunnel construction; *II* after tunnel construction; *III* after the tunnel reconstruction

is dependent upon various tunnel capacities. Figure 9.11 shows that although the reconstruction of the 1974 tunnel would reduce the volume of floods in the polje to 30×10^6 m^3, it would significantly decrease the flood level, i.e. it would be decreased by only 120 cm. Two additional alternatives were examined (hydrographs 3 and 4) for different tunnel capacities. Floods still appear, but their duration is reduced and the intensity limited. It has been definitely proved that it is not economically justifiable to increase the tunnel capacity beyond $45-50$ m^3 s^{-1} since extensive works would have to be carried out on the regulation of the Matica Vrgorska River. Even then, floods would not be eliminated, but only reduced. Therefore, it is better from an economic standpoint to permit the occasional flooding of the lowest polje section since the entire elimination of flooding would entail high costs.

As displayed in Figure 9.9 the water from the Vrgorsko Polje flows through the tunnel to the Baćinska Lakes. The hydrologic catchment area of these lakes is 21.4 km^2. Their hydrologic regime has been significantly changed as great quantities of water were delivered there from the Vrgorsko Polje through the tunnel. In 1913 a tunnel was drilled leading from the Baćinska Lakes to the Adriatic Sea. Before it was constructed the water from the Baćinska Lakes slowly flowed to the Adriatic Sea through a system of karst cracks (channels, fissures). The water levels in the lakes rearches upto 7.5 m a.s.l. due to the slow emptying of water. After the construction of the tunnel the highest water levels recorded on the lakes have not exceeded 3.5 m a.s.l. The hydrologic analysis was carried out for the three periods (Bonacci and Švonja 1984): (1) 1926–1938; (2) 1940–1958; and (3) 1975–1981. The water level regime of the lakes was analyzed and the selection of the periods conditioned by the phase of the construction of the drainage structures. In the first period there existed only the Baćinska Lakes–

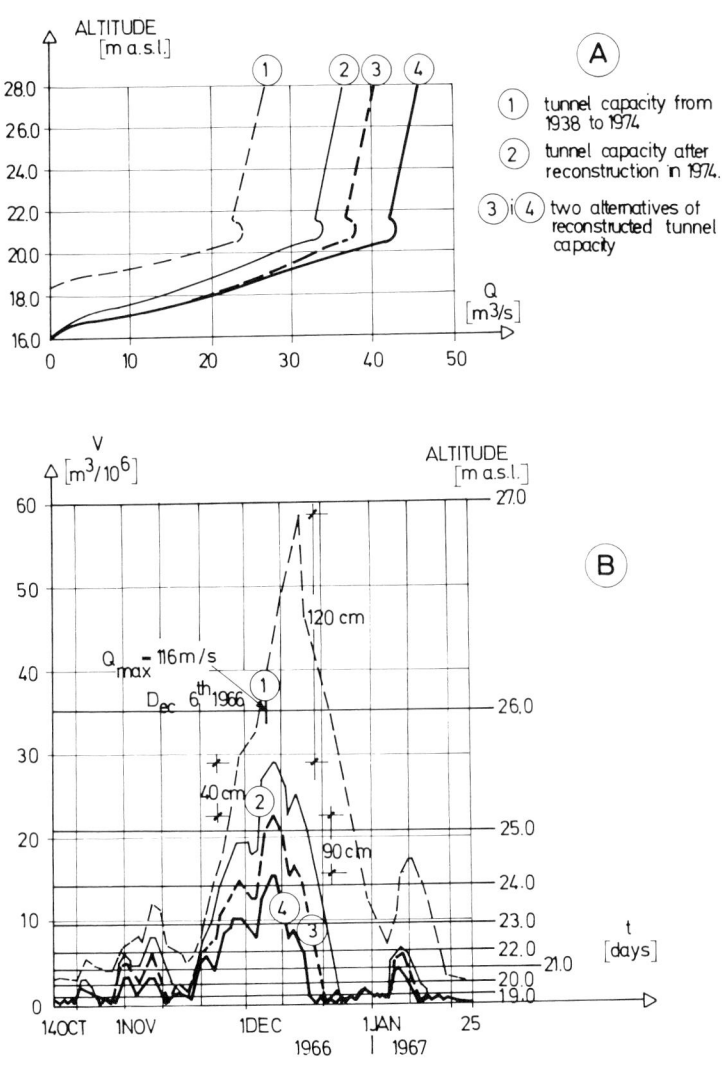

Increase in the Capacity of the Outlet Structures 163

Fig. 9.11 A, B. Discharge curves of the tunnel in different periods (**A**) and water levels versus time of a flood in the Vrgorac Polje (Yugoslavia) depending on the tunnel capacity (**B**)

Adriatic Sea Tunnel. In the second period, the Vrgorsko Polje – Baćinska Lakes Tunnel was in operation and the tunnel leading to the sea was widened, whereas in the third period this tunnel was reconstructed. The hydrographs of the first period are uniform and show that the water regime is natural. The maximum water levels occurred at the beginning of winter, and the minimum at the end of summer. The highest measured water level of the lakes was 2.63 m a.s.l. The water levels recorded in the second period show that the water waves in this period were higher and lasted for a shorter period of time than in the preceding period because of the emptying of water. The oscillations of the wave peaks (generally

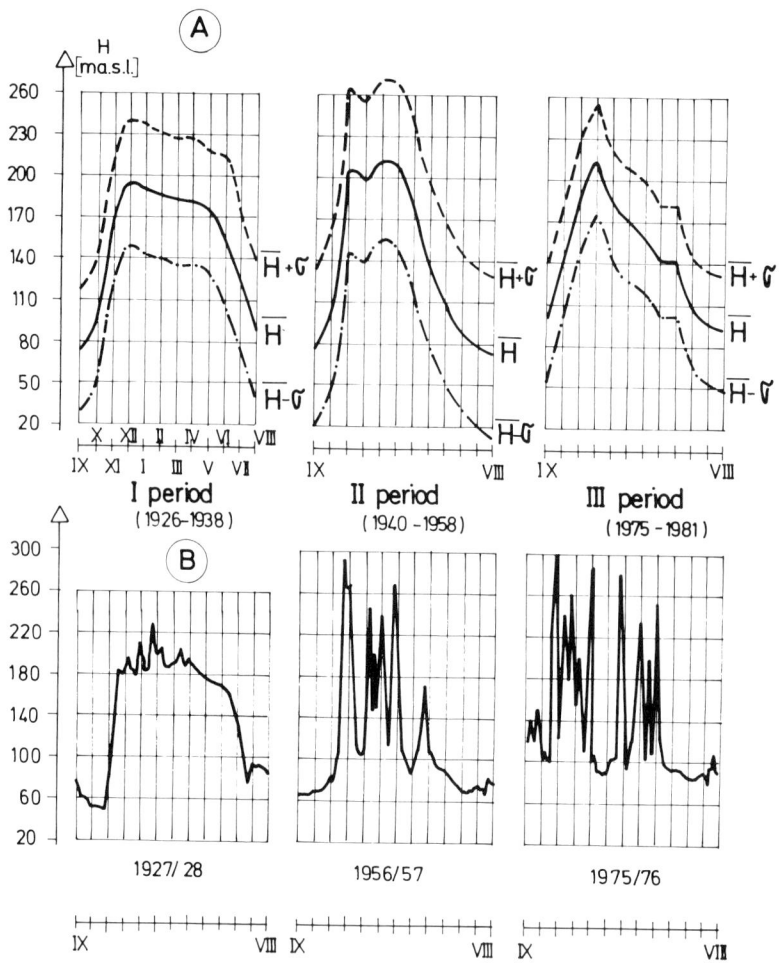

Fig. 9.12 A, B. Water level of the Bačina Lake (Yugoslavia) in three analysed periods. **A** Mean water level hydrographs in three periods; **B** water level hydrographs in three hydrologic years of each analysed period

four waves a year) occurred more frequently, generally from October to March, when the inflow from the Vrgorsko Polje was the greatest. The maximum recorded water level in this period was 3.18 m a.s.l. In the third period the water waves were the highest, of shortest duration and had the greatest number of peak oscillations (ten a year, on the average). Since precipitation in the Vrgorsko Polje and at higher horizons is heaviest from October to March, the inflow through the reconstructed inflow tunnel is the greatest in that period, and the water levels of the lake the highest with a maximum value of 3.28 m a.s.l. Figure 9.12 presents the mean hydrographs over several years and one typical yearly hydrograph for each of the analyzed periods. Evidently, the changes in the dimensions of the outlet structures affected the changes in the hydrologic regime. It has become

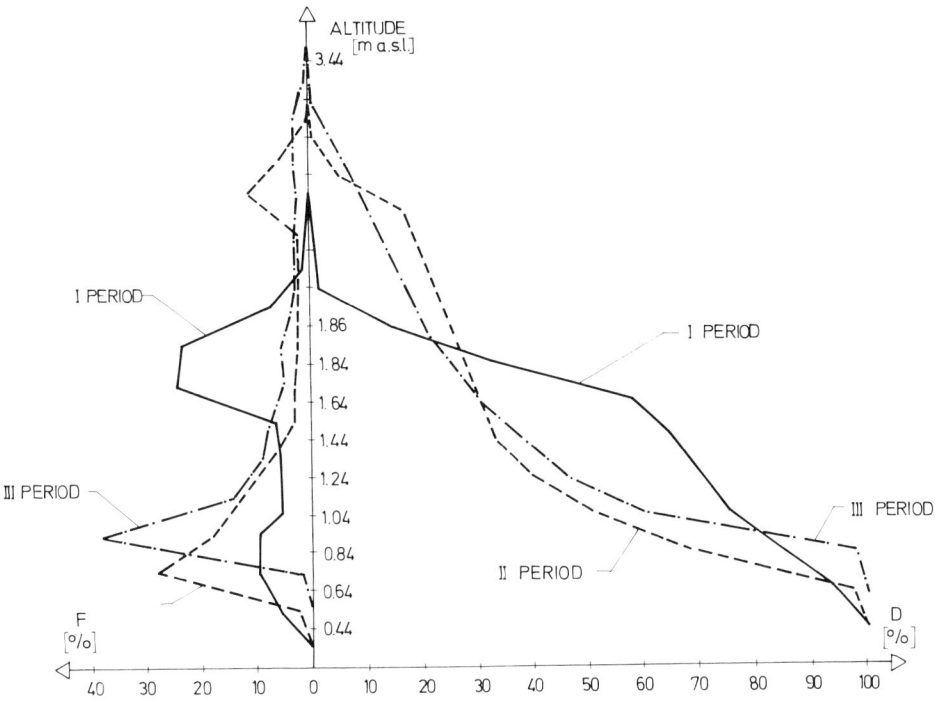

Fig. 9.13. Water level duration and frequency curves for three periods on the Bačina Lakes (Yugoslavia)

more dynamic with more fluctuations. Figure 9.13 shows that there are certain changes in the shape of the frequency and duration curves. The first period significantly differs, whereas there are no great differences in the second and third periods. The construction of the tunnel significantly changed the regime of the waters of the downstream horizon.

The problems of changing the natural conditions of water evacuation from the polje are very complex and the civil engineering and economic aspects as well as the changes in the hydrologic regime should be considered. The decrease of the water level and shorter duration lead to worsening conditions of the water regime on the downstream horizon. In order to eliminate this problem a great deal of regulation and protection work has been completed in the downstream poljes with less extensive work in the upstream poljes. The capacity of the outlets should often be increased. Figure 9.14 gives a schematic presentation of two small karst poljes under similar conditions. An excavation from Polje *a* to Polje *b* was made to provide free outflow of water from the upper polje, which is flooded an average of 8 months a year, since its ponor's swallow capacity is insignificant. An excavation with a lock was provided to regulate the outflow from Polje *a* to *b* and to prevent its being flooded. In order to ensure the outflow from Polje *b* to lower horizons, a tunnel was drilled. All those structures were built 20 years ago. As agriculture was intensified it became necessary to entirely eliminate flooding even of short duration. It should be noted that the soil in karst poljes is very fertile

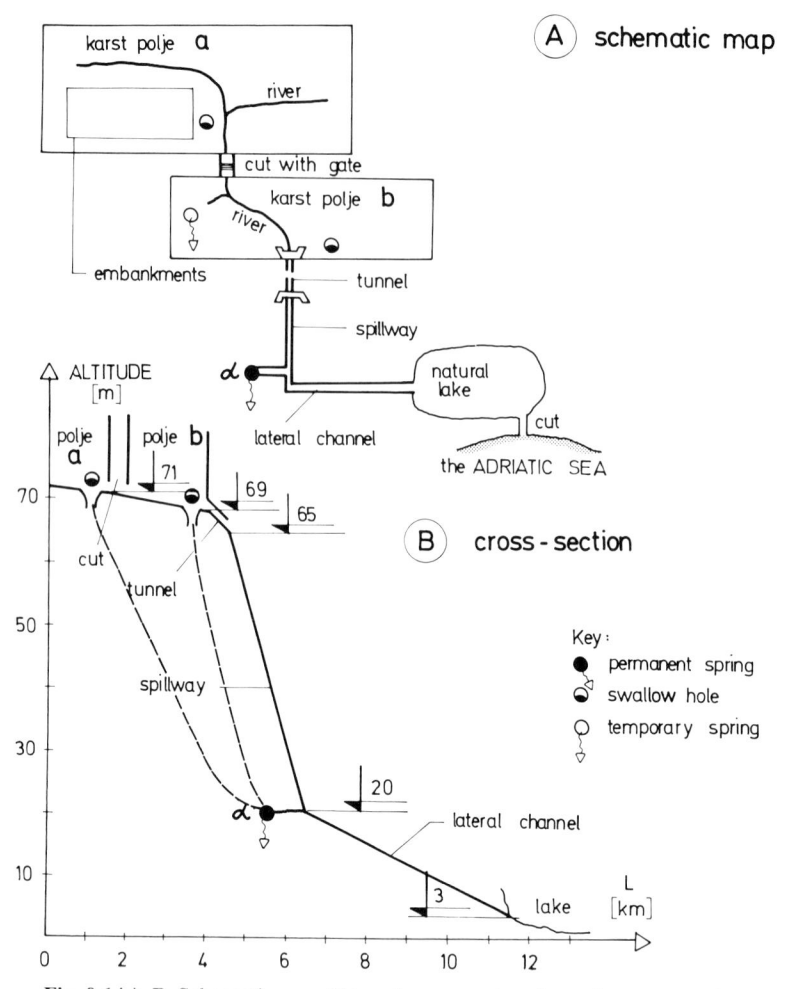

Fig. 9.14 A, B. Schematic map (A) and cross-section through a system of two poljes in the karst (B) (Yugoslavia)

and the climatologic conditions make it possible to have several harvests and to raise Mediterranean strands. Consequently, the hydrologic regime in the previously mentioned system of the polje was considered repeatedly in order to define new and more extensive work on tunnel drillings and excavations. Different combinations of the outlet capacities were studied and for each combination the number of hours when the water was outside the regulated riverbeds, i.e. when it floods the arable areas. Floods with different return periods have been considered. According to hydrologic and economic analyses a decision was made. It was necessary to increase the tunnel capacity by 100%. A plan for optimal management of a lock and excavation was developed under the function of water stage conditions in Poljes *a* and *b*.

9.3 Surface Hydrotechnical Works

The term "surface hydrotechnical works" means the regulation works on open streams and the channel construction for surface drainage as well as large land reclamation and other measures. It can be superficially assumed that surface drainage cannot have a considerable influence on the change of the underground water level which essentially governs the outflow processes in karst. This is not always the case. Such works are not completely harmless nor without consequences. This has been illustrated in the case of the Clarinbridge River Catchment in Ireland described by Drew (1984). A surface drainage network has been considerably developed (Fig. 9.15). It caused some consequences which were unforeseeable, but which, depending upon the geology, could be more or less expected in other karst regions. The following effects have been noted: (1) the high density of the surface outflow network increased the direct outflow; therefore, underground sinking decreased. Underground water that had previously appeared along the entire catchment area was located due to its fast transportation along the sink lines which depend upon the season; (2) surface channels bored in limestone made faster and more direct water penetration possible, but created underground pollution; (3) natural protection was broken on channel lines which resulted in many environmental problems; (4) essentially, the reserves of ground-

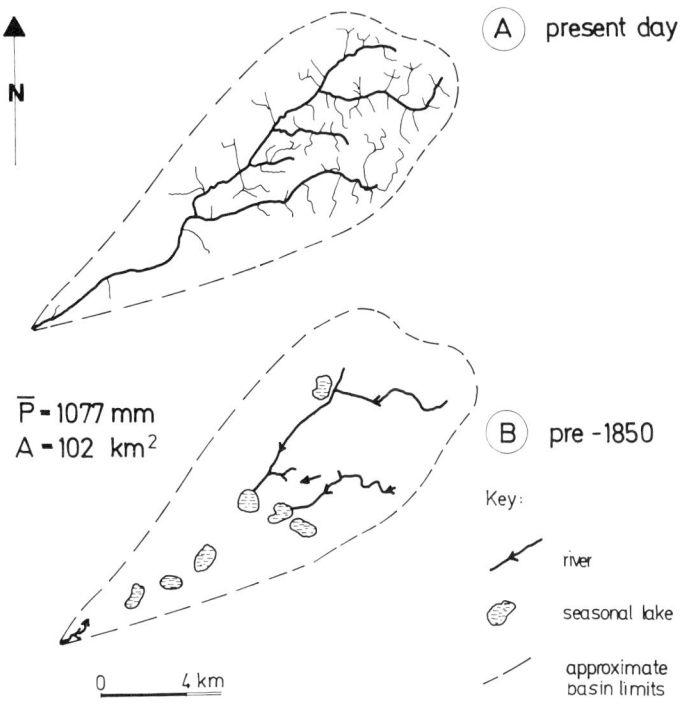

Fig. 9.15 A, B. Surface drainage in Clarinbridge (Great Britain) at (**A**) present-day and (**B**) pre-1850. (Drew 1984)

water were decreased and were partly polluted which has led to economic and other consequences.

In this case the works lasted longer covering a period of several tens of years with the consequences occurring gradually and slowly. Similarly, it is possible to consider the influence exerted by planting or cutting trees on the change in the hydrologic regime of catchments in karst. Surface hydrotechnical works involve also the opening of quarries or the construction of road or railroad networks. The construction of these structures significantly affects the subcutaneous zone and hence the processes of water circulation in karst. Therefore, when designing these structures in karst attempts should be made to avoid and eliminate the negative consequences and the delayed hazards.

9.4 Action on the Groundwater

The term "action on the groundwater" refers to the pumping of the groundwater from karst, the disposal of waste water and liquid wastes into the limestone aquifer and the artificial recharge of karst groundwater.

The pumping of the groundwater from karst for irrigation purposes and for water supply is a common phenomenon. Moreover, today, it is gaining prominence as a method of water supply under karst conditions. When the groundwater is abruptly and inadequately pumped out, it results in fractures of the upper layers in unstable terrains because of fissures in the karst mass which are mutually well linked. This can result in the formation of new ponors and can activate those in existence. The pumping of the groundwater in karst also significantly lowers the groundwater level in its path. Hall and Metcalfe (1984) gave as an example the collapse of 22 ponors caused by pumping groundwater in the vicinity of Dover, Florida (USA) in January 1977, which caused extensive damage.

Consequently, when pumping the groundwater from a karst aquifer, and particularly in densely populated areas, it is necessary to carefully measure and assess the pumping effects and to consider them in the design of the structures. The quantities of pumping are not restricted exclusively by the capacity of the groundwater, but also by the stability of the geologic structures. Considering the procedure of water intakes carried out by pumping the well, it should be stressed that the water inflow into the well originates in the slowly inflowing groundwater storages, since the flow in the shaft system of karst is quite rapid. Thus, the water is retained for only a short time and cannot be exploited as a long-lasting, reliable and efficient water supply.

During the last 10 years, the waste water and the liquid wastes have been more intensively disposed into the deeper layers of the karst aquifers. Meyer (1984) referred to the disposal of partially treated and untreated municipal waste water into deep wells by injection. In 1969, a test well was drilled near Miami, Florida (USA), to a depth of 1000 m in order to determine whether or not the hydrogeologic conditions were favourable for the successful injection and storage of the effluent. The hydraulic characteristics of the zone to be used for the disposal of waste water were assessed after expensive investigations. It was concluded that the zone was capable of accepting high rates of liquid injection in spite of its relatively low permeability. The potential for upward irrigation by buoyant forces has been recognized by the regulative agencies and the monitoring requirements have been established to assess the potential.

Vecchioli et al. (1984) investigated the alternations in chemical composition caused by the injection of industrial liquid waste containing organonitrate compounds and nitrate ions into the lower limestone layers of the Florida aquifer at a site near Pensacola, Florida (USA). The depth of the injection wells was ca. 450 m. In addition to the injection wells, several monitor wells were also drilled. Injection of the industrial liquid waste began in June 1975. According to long-lasting, complex and expensive investigations, it was concluded that some positive chemical changes in the underground layers occurred.

Morozov and Lushchik (1984) reported on experimental investigations of the artificial groundwater recharge in two karst regions in the USSR. The investigations have shown that the artificial recharge of the karst water is fairly effective provided that the raw water is decontaminated and the recharge installations are operated continuously. It is necessary to carry out extensive and detailed investigations related to various scientific disciplines and fields regardless of which activity is involved, i.e. either pumping the water from a karst aquifer, its recharge or the disposal of the waste water. Any action on the groundwater can cause very serious negative consequences on the water quality and quantity. Therefore, these activities should be carried out with utmost attention and care and the investigations should be at a high professional and scientific level.

9.5 Usage of the Karst Spring Water

The modern trend in the exploitation of the karst springs water is to use a great part of the static reserves of the groundwater belonging to a certain spring. The groundwater in karst can be used for water supply and this is often practiced. Various technical procedures can be used to increase the capacity of the karst springs by entering the zone of the groundwater static storage. Three different cases are presented in Figure 9.16. Figure 9.16A presents the increase in the spring capacity by drilling a tunnel (horizontal intake) or by constructing a gourd for siphoning. Bagarić (1981) reported on the construction of a gourd at the Radobolja Spring (Yugoslavia), and Kullman (1984) on the intake of the Biele Vodỳ (Czechoslovakia), by boring a tunnel (Fig. 9.17A). Figure 9.17B presents the difference between the hydrographs of the natural Spring 1 and the new intake of Spring 2, which is significant and amounts to $37 l s^{-1}$ or even 40% more than the natural state. Figure 9.16B presents one possibility of increasing the water storage in the karst underground. Placing the concrete cork in Position 1 at the spring (Bagarić 1981) can bring about the formation of sliding and rockslides of the slope above the spring. Consequently, the cork should be placed in Position 2 to avoid this damage. It does not cause a significant decrease in the recently formed groundwater storage. Figure 9.16C presents another possibility for increasing the spring capacity; this time by intake using the well and the pumps. The Lez Spring near Montpellier (France) was taken in using this method (Drogue 1983) and its discharge in the period of minimum water quantities was increased

Fig. 9.16 A–C. Three possibilities to increase the capacity of karst springs by (**A**) drilling a tunnel or by constructing a gourd of siphoning; (**B**) placing the concrete cork on the exit; (**C**) using the pumps

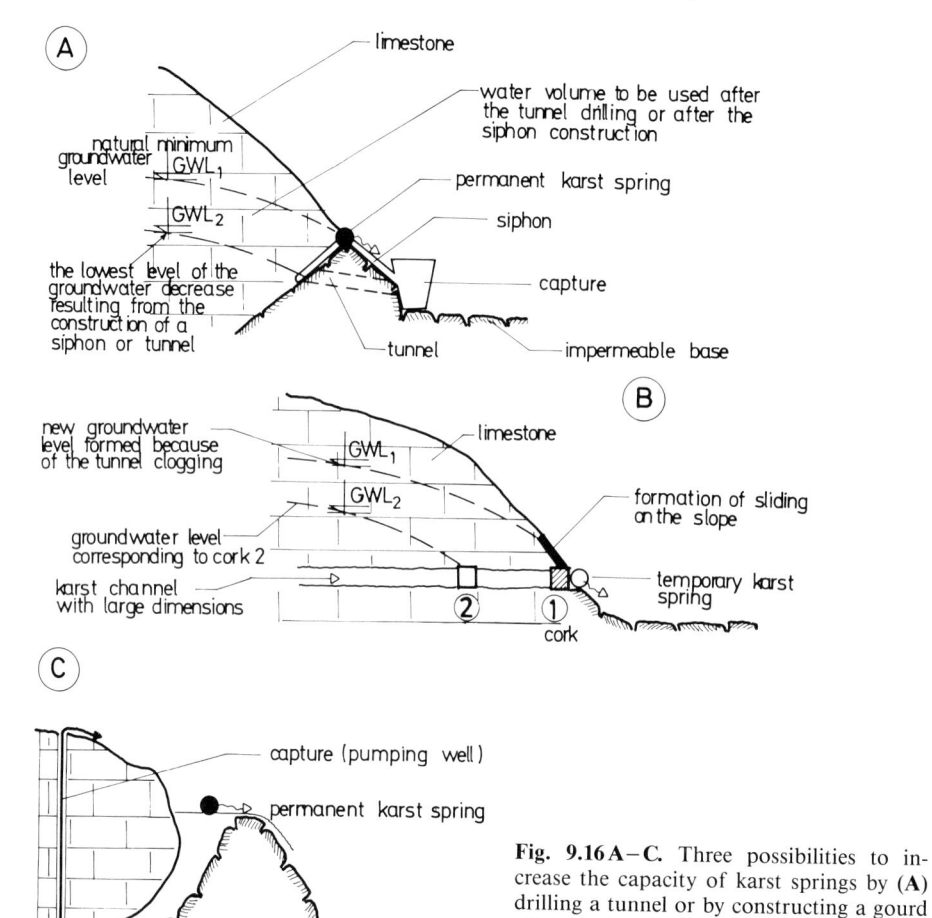

from $400 \, l \, s^{-1}$ under natural conditions to a maximum quantity of $1700 \, l \, s^{-1}$ by locating the pump 35 m below the level of the spring spillway. These data alone sufficiently stress the dimensions of their use, since in the natural state that quantity of water is practically unexploitable. One of the possibilities for water exploitation in karst is to drill vertical wells in the hinterland and to locate the pumps deep into the zone of the static reserves of the groundwater. The intake of groundwater from the hinterland of the springs makes it possible to ensure a better quality of drinking water. The maximum water quantities can be used and the most reliable protection measures can be provided only by a detailed study of the mechanism of the spring functioning and the connections between the fissure systems. Consequently, extensive investigations related to these problems should be effected.

When there is a cave in the hinterland of the ascendent karst spring of the lake type, it is possible to influence that change in the spring capacity by constructing appropriate works. Figure 9.18 is a schematic representation of the functioning

The Development of the Brackish Karst Springs 171

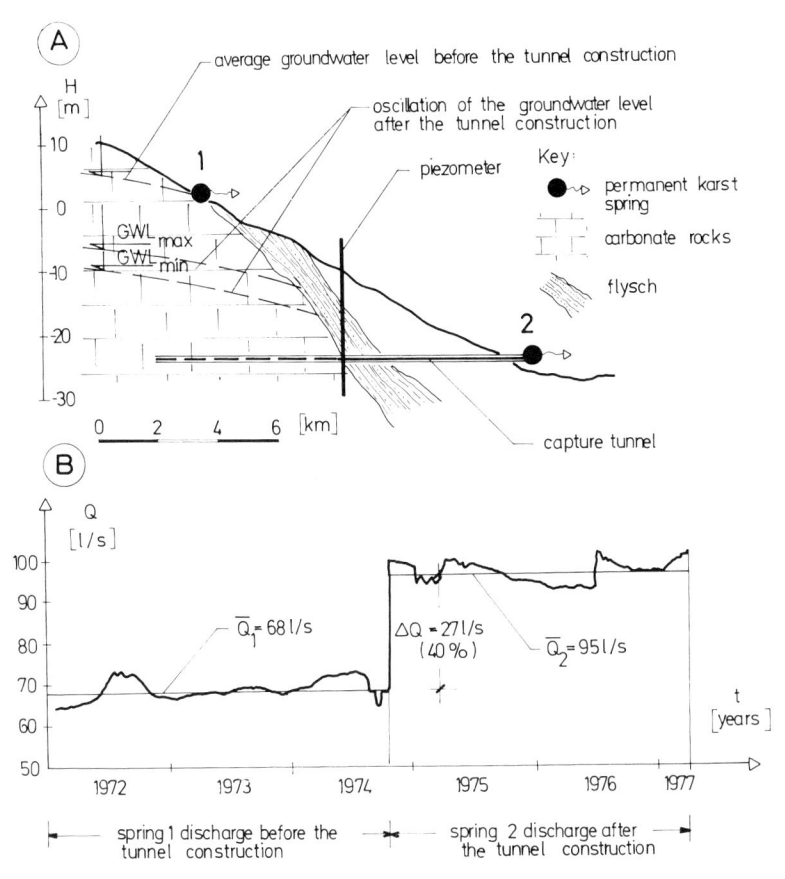

Fig. 9.17 A, B. Capture of the karst spring illustrated on the example of the Biele Vody (Great Tatra – Czechoslovakia). **A** Cross-section through spring; **B** discharge hydrographs for different conditions (Kullman 1984)

system of the previously described karst spring. The spring capacity can be influenced in two ways: by the erection of a dam with locks in front of the spring and by the formation of a small surface reservoir (level 3 in Fig. 9.18). In the low water period the spring capacity can be increased by lowering the water level in the surface reservoir by means of a lock. In the flood period when the spring discharges are abundant, it is possible to increase the dynamic reserves of the groundwater by raising the water level in the surface reservoir.

9.6 The Development of the Brackish Karst Springs

Large quantities of spring water in the karst coastal areas are brackish and cannot be used either as potable (drinking) water or for industrial purpose. The greatest demand for water in those regions occurs in the dry summer period, particularly in those places oriented towards tourism. Although the spring capacity in that

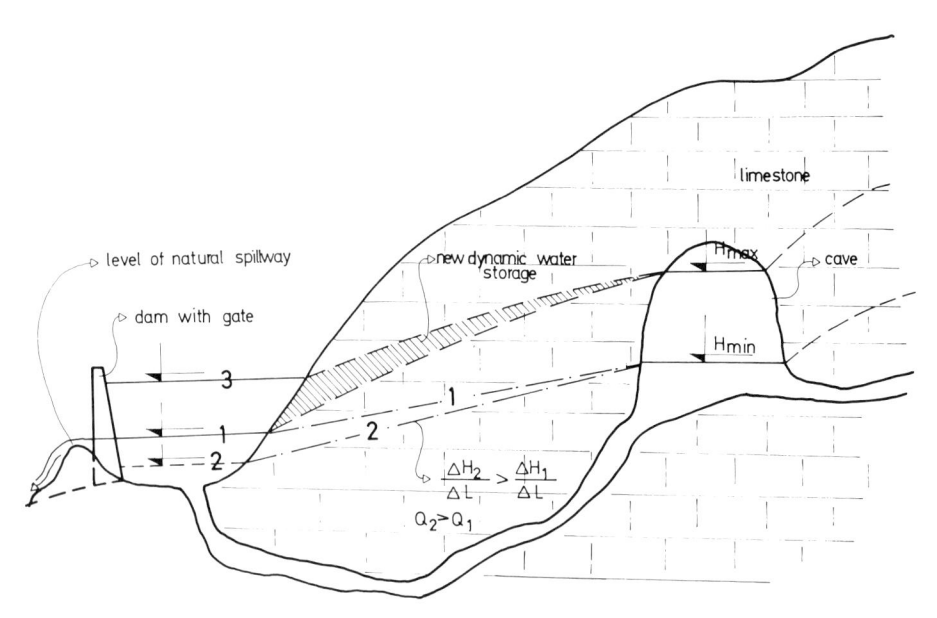

Fig. 9.18. Possibility of outflow regulation from a special type of karst spring

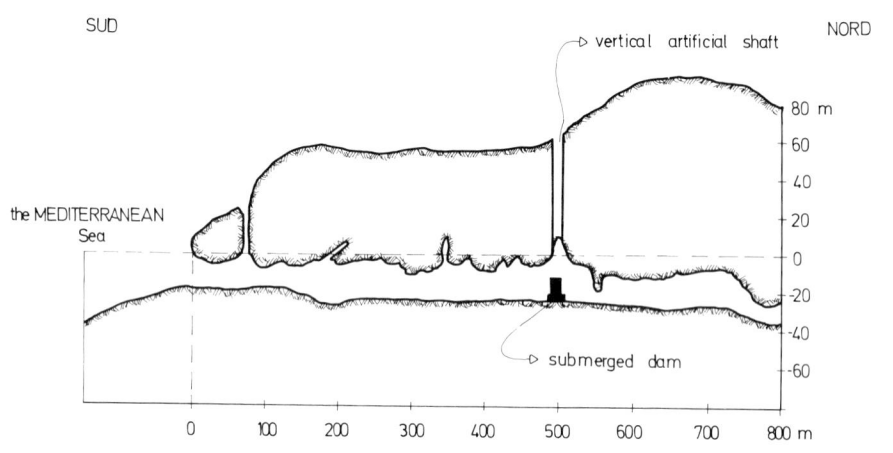

Fig. 9.19. Cross-section through a karst channel near Port-Miou (France). (Potie 1973)

period is reduced, the water quantities are most frequently more than sufficient for satisfying the existing needs. The salinity of water, however, is highest in those periods. Consequently, since ancient times, man has attempted to desalinize the water of the coastal springs and to exploit the enormous reserves of fresh water found in the vicinity of the seacoast. This task has been quite complex. Very often these attempts have proven to be unsuccessful, although significant financial means have been invested in the investigation and studies related to these activities.

The interception of fresh water above the zone of mixing is the most suitable method of development since the fresh and saline water are mixed within the karst mass, relatively far from the seacoast. This mixing can be prevented in various ways, depending upon the type and characteristics of the brackish karst springs. The groundwater of the coastal karst aquifers of isotropic permeability in Israel was intercepted by a line of wells. The construction of a dam and the isolation of the mouth from the sea water can be useful for those springs contaminated at the mouth. The Ayios Georegios Spring (Greece) and Sorgenti d'Aurisina Spring near Trieste (Italy) were formed by the construction of dams (Breznik 1978). Several successful intakes of fresh water via tunnels of great length have been carried out along the coast of the Adriatic Sea in Yugoslavia (Mijatović 1984a, b). One tunnel was built 2 km from the coast (the Roman Well) and was 250 m long. The tunnel in Herceg Novi was much longer and its length was ca. 1000 m. The tunnels are used as drainage systems for collecting the fresh water before it is mixed with the sea water. They should be drilled within the karst mass perpendicular to the direction of the general flow of fresh water flowing towards the coastal springs. A large tunnel was constructed in the hinterland of the Pantan Spring (Yugoslavia), but its attempt at fresh water intake has not been successful.

Figure 9.19 displays a cross-section of the karst mass near Port-Miou (France) along the coast of the Mediterranean Sea. An underground karst river flows along a large karst channel lying 20 to 40 m below sea level. Since the river water mixes with the sea water, it could not be used for drinking. The construction of the submarine dam about 10 m high prevented the salizination of the water upstream from the dam. Consequently, the river water can be used for the water supply of the region (Potie 1973).

References

Alfirević S (1966) Les sources sous-marines de la baie de Kaštela. Acta Adriat 6(12):1–38
Alfirević S (1969) Adriatic submarine springs in the water system of the Dinaric karst littoral and their problems. Carsus Iugoslaviae 6:183–205
Alfirević S (1970) Les aspects hydrogéologiques de la circulation des eaux souterraines sur la côte orientale de la mer. Adriatique 15–29
Atkinson TC (1977) Diffuse flow and conduit flow in limestone terrain in the Mendip Hills, Somerset (Great Britain). J Hydrol 35:93–100
Avdagić I (1976) Determination of flow through flooded karst poljes by use of poljes' and piezometric boreholes' water levels. Proc Karst hydrology and water resources. WRP 341–354
Babushkin VD, Lebedyanskaya ZP, Plotnikov II (1984) The distinctive features of predicting total water discharge to deep-level mines in fissure and karst rocks (Exemplified by one of the northern Urals mineral deposits). In: Burger A, Dubertret L (eds) Hydrogeology of karstic terrains. 1:229–232
Bagarić I (1981) Određivanje veličine dinamičkog dijela prirodne akumulacije vode (Size defining of the dynamic partion of natural water storage). Naš Krš 6(10–11):187–204
Balkov VA (1967) Vlijanie karsta na minimaljnij stok (Karst influence on minimum flow). Proc Moscow Univ Seminar Perennial variations of runoff and the stochastic methods of its estimation. pp 201–207
Barbalić Z (1976) Properties of water resource systems of enclosed karst plains. Proc Karst hydrology and water resources. WRP 815–828
Bögli A (1980) Karst hydrology and physical speleology. Springer, Berlin Heidelberg New York
Bojanić L, Ivičić D, Božičević S (1980) Inženjersko geološka problematika dovodnog tunela HE Zakučac II (Engineering geology explorations for the HP Zakučac II tunnel). Proc 6th Yug Symp Hydrogeology and Eng Geology, Book 2, pp 191–200
Bonacci O (1979) Influence of turbulence on the accuracy of discharge measurements in natural streamflows. J Hydrol 42:347–367
Bonacci O (1981) Establishment of quantities of water sinking on karst sections of streams. Naš krš 6,10–11:139–151
Bonacci O (1982) Specific hydrometry of karst regions. Proc Symp Advances in Hydrometry, Exeter. IAHS Publ 134:321–333
Bonacci O (1985a) Hydrological investigations of Dinaric karst at the Krčić catchment and river Krka springs (Yugoslavia). J Hydrol 82:317–326
Bonacci O (1985b) Flooding of the poljes in karst. Proc 2nd Int Conf Hydraulics of floods and flood control, Cambridge, Sept 24th–26th, pp 119–136
Bonacci O, Perger V (1970) Utvrđivanje gubitaka voda kod stagnirajućih vodostaja na potezima kraških vodotoka (Evaluation of water losses on a river in the karst). Gradevinar 22(5):155–164
Bonacci O, Roglić S (1982) Application of discrete dynamic programming for overland flow analysis, vol 2. Proc 1st Int Sem Urban Drainage Systems, Southampton, pp 165–177
Bonacci O, Švonja M (1984) Promjena vodnog režima Bačinskih jezera (Change of water regime of the Bačina Lake). Gradevinar 36(2):53–58
Borelli M (1966) O gubicima vode iz kraških akumulacija, Akumulacija Buško Blato. (Water losses from karst reservoirs, Buško Blato Reservoirs). Saopštenja Inst J Černi 36:17–30
Borelli M, Pavlin B (1967) Approach to the problem of the underground water leakage from the storages in the karst regions, vol 1. Proc Dubrovnik Symp Hydrology of fractured rocks, pp 32–62

Borić M (1980) Use of natural thermal properties of water in karst storage for locating water leakages – Case of storage Buško Blato, Zbornik radova 6. Jug Simpoz o Hidrogeologiji i Inž Geol, Knjiga 2:179–190

Božičević S, Pavičić A, Renić A (1983) Pojava vodnog toka u spiljskom sistemu ispod sedrene barijere (Occurrence of water streamflow in cave below waterfall). Proc of Yug Symp of Hydrology, Split, pp 251–260

Breznik M (1973) Nastanke zaslanjenih kraških izvirov in njihova sanacija (The origin of brackish karstic springs and their development). Geol – Razprave in Poročila 16:83–186

Breznik M (1978) Mechanism and development of the brackish karstic spring Almyros Iraklion. Ann Géol Pays Helléniques, pp 29–46

Breznik M (1983) Večnamenska akumulacija Cerkniško jezero (The cerknica lake multi-purpose storage). Gradbeni Vestnik 1–2:3–15

Brook GA, Ford DC (1980) Hydrology of the Nahanni Karst, northern Canada, and the importance of extreme summer storms. J Hydrol 46:103–121

Brown MC, Ford DC (1971) Quantitative tracer methods for investigation of karst hydrological sytem. Trans Cave Res Group GB 13(1):37–51

Brucker RW, Hess JW, White WB (1972) Role of vertical shafts in the movement of ground water in carbonate aquifers. Ground Water 10(6):9

Burger A, Pasquier F (1984) Prospection at captage d'eau par forages dans la vallée de la Brevine (Jura Suisse). In: Burger A, Dubertret L (eds) Hydrology of karstic terrains, vol. 1 UNESCO, 145–149

Coutagne A (1949) L'évaporation des surfaces d'eau naturelles. Rev Gen Hydraulique 7–8:174–184

Cowell DW, Ford DC (1983) Karst hydrology of the Bruce Peninsula, Ontario, Canada, In: Back W, LaMoreaux PE (Guest-Eds) VT Stringfield symp – processes in karst hydrology. J Hydrol 61:163–168

Čorović A, Cerić A, Milina J, Tomić M (1985) Pristup rješavanju zaštite kraških vrela (An approach to the protection of karst springs). Vodoprivreda 17, 96–97:200–205

Cvijić J (1983) Des Karstphänomen. Pencks Geogr Abh V3:217–330

De Vera MR (1984) Rainfall-runoff relationship of some catchments with karstic geomorphology under arid to semi-arid conditions. In: Stout GE, Davis GH (eds) Global water: science and engineering. The Ven to Chow memorial volume. J Hydrol 68:85–93

Dreiss SJ (1983) Linear unit-response function as indicators of recharge areas for large karst springs. In: Back W, La Moreaux PE (Guest-Eds) VT Stringfield symp – processes in karst hydrology. J Hydrol 61:31–44

Drew D (1984) The effect of human activity on a lowland karst aquifer. In: Burger A, Dubertret L (eds) Hydrogeology of karstic terrains, vol 1. UNESCO, pp 195–199

Drogue C (1972) Analyse statistique des hydrogrammes de décrues des sources karstiques. J Hydrol 15:49–68

Drogue C (1980) Essai d'identification d'un type de structure de magasins carbonates, fissurés. Mem H Ser Soc Geol (France) 11:101–108

Drogue C, Laty AM, Paloc H (1983) Les eaux souterraines des karsts méditerranéens. Exemple de la région pyrénéo-provençale. Bull BRGM, Hydrogéol Géol Ing 4:293–311

Dukić D (1984) Hidrologija kopna (Land hydrology). Naučna Knjiga, Beograd, p 498

Ede DP (1973) Aspects of karst hydrology in South Wales, part 2. Gewogr J 139:28–294

Forkasiewicz J, Paloc H (1967) La régime de tarissement de la Foax de la Vis, vol 1. Dubrovnik Symp. Hydrology of the fractured rocks, pp 213–226

Foster SSD, Milton VA (1974) The permeability and storage of an unconfined chalk aquifer. Hydrol Sci Bull 19:485–500

Fritz F (1978) Hidrogeologija Ravnih Kotara i Bukovice (Hydrogeology of Ravni Kotari and Bukovica). Carsus Iugoslaviae 10/1, JAZU Zagreb: 1–43

Fritz F, Pavičić A (1982) Hidrogeološki viseći dijelovi rijeke Krke i Zrmanje (Hydrogeological investigations of karst sections of the Krka and Zrmanja Rivers). Proc 7th Yug Symp Hydrogeology and Eng Geol, Book 1, pp 115–121

Gale SJ (1984) The hydraulics of conduit flow in carbonate aquifers, J Hydrol 70:309–327

Gavazzi A (1904) Die Seen des Karstes, vol 2. Abh Geogr Ges Wien

Gavrilov AM (1967) On the problem of the influence of karst on the hydrological regime of rivers, vol 1. Proc of the Dubrovnik Symp Hydrology of Fractured Rocks, pp 544–562

References

Gavrilović D (1967) Intermitentni izvori u Jugoslaviji (Intermittent springs in Yugoslavia). Glas Srp Geogr Društva 47(1):125–139

Ghyben WB (1889) Nota in verband met de voorgenomen putboring nabij Amsterdam (Notes on the probable results of the proposed well drilling near Amsterdam). Inst Ing Tijdchr 21

Gjurašin K (1942) Prilog hidrografiji primorskog krša. (Contribution to the litoral karst hydrography). Tehnički Vjesnik 59:107–112

Gjurašin K (1943) Prilog hidrografiji primorskog krša. (Contribution to the litoral karst hydrography), Tehnički Vjesnik 60:1–17

Glanz T (1965) Das Phänomen der Meermühlen von Argostolion. Steir Beitr Hydrogeol 17:113–128

Gospodarič R (1976) Razvoj jam med Pivško Kotlino in Planinskim poljem v kvartarju. Acta Carsol 7:5–139

Gospodarič R, Habič P (1978) Kraški pojavi Cerkniškega polja (Karst phenomena of Cerkniško Polje). Acta Carsol VIII/1:1–162

Grund A (1903) Die Karsthydrographie. Pencks Geogr Abh Leipzig, 7:3

Gunn J (1982) Point-recharge of limestone aquifers – a model from New Zealand karst. J Hydrol 61:19–29

Habič P (1982) Kraški izvir Mrzlek, njegovo zaledje in varovalno območje (Mrzlek karst spring, its catchment and protection area). Acta Carsol 10:45–73

Hajdin G, Avdagić I (1982) Prilozi za hidrauliku krša (Contributions to karst hydraulics). Proc Yug Symp on Hydraulic, Portorož, pp 1–13

Hajdin G, Ivetić M (1978) Jedan primjer pokušaja objašnjavanja uvjeta u kraškom podzemnom toku na osnovi opažanja piezometarskih stanja i izlaznih proticaja (An attempt to explain the hydraulic conditions in karst underground circulation according to piezometer measurements and outflow discharges). Naš Krš 4(4):27–32

Hall LE, Mecalfe SJ (1984) Sinkhole collapse due to groundwater pumpage for freeze protection irrigation near Dover, Florida, January 1977. In: Burger A, Dubertret L (eds) Hydrogeology of karstic terrains, vol 1. UNESCO, pp 248–251

Herzberg B (1901) Die Wasserversorgung einiger Nordseebäder, vol 4. Gasbeleuchtung, Wasserversorgung, München, p 4

Hubbert MK (1940) The theory of groundwater motion. J Geol USA 8:48

Ivanković T (1976) Hydrogeologic estimation of groundwater storage connected with the surface water storage. Proc Karst hydrology and water resources. WRP 193–206

Johnston RH (1983) The saltwater-freshwater interface in the tertiary limestone aquifer Southeast Atlantic outer-continental shelf of the USA. In: Back W, LaMoreaux PE (guest-eds) VT Stringfield Symp. – processes in karst hydrology. J Hydrol 61:239–249

Karanjac J, Altug A (1980) Karstic spring recession hydrograph and water temperature analysis: Oymapinar Dam project, Turkey. J Hydrol 45:203–217

Karanjac J, Günay G (1980) Dumanly Spring, Turkey – the largest karstic spring in the world. J Hydrol 45:219–231

Kayane I, Taniguchi M, Sanjo K (1985) Alteration of the groundwater thermal regime caused by advection. Hydrol Sci J 30, 3, 9:343–360

Knežević B (1962) Hidraulički problemi karsta (Hydraulic problems of karst). Saopštenja Instituta J Černi, (Beograd), 25:1–14

Knežević B, Voinović M (1962) Laboratorijska ispitivanja shema podzemith tokova u kršu (Laboratory investigations of karst groundwater connections). Proc 3rd Yug Symp Hydraulic, Opatija, pp 48–53

Kogovšek J (1982) Vertikalno prenikanje v Planinski jami v obdobju 1980/81. (Vertical percolation in the Planina Cave). Acta Carstol 10:107–125

Kogovšek J, Habič P (1980) Preučavanje vertikalneg prenikanja vode na primerah Planinske in Postojnske jame. (The study of vertical water percolation in the case of Planina and Postojna Caves). Acta Carsol 9:133–148

Kohout FA (1966) Submarine springs. A neglected phenomenon of coastal hydrology. Proc Symp on hydrol and water resources, Ankara

Komatina M (1984) Hidrogeološka istraživanja – Metode istraživanja (Hydrogeological investigations – Methods of investigations). Geozavod, Beograd, Yugoslavia, p 375

Koudelin BI, Karpova VP (1967) The effect of karst on the regularities of groundwater formation, vol 1. Porc Dubrovnik Symp Hydrology of Fractured Rocks, pp 208–212

Kullman E (1984) Captage d'une source karstique par forages horizontaux – Example de la source Biele Vodỳ. In: Burger A, Dubertret L (eds) Hydrogeology of karstic terrains, vol 1. UNESCO, pp 123–125

Kuščer I (1950) Kraški izviri ob morski obali (Litoral karst springs). Razprave SAZU, Ljubljana, pp 99–137

LeGrand H (1983) Perspective on karst hydrology. In: Back W, La Moreaux PE (Guest-Eds) VT Stringfield Symp – Processes in karst hydrology. J Hydrol 61:343–355

LeGrand H, Stringfield VT (1973) Karst hydrogeology – a review. J Hydrol 20:97–120

Lehmann O (1932) Die Hydrographie des Karstes. Enz Erdkd, 6b, Leipzig-Wien

Magdalenić A, Jurak V, Bonacci O (1986) Analysis of a karst spring (Yugoslavia), Karst water resources, Ankara, July 1986. IAHS Publ 161:359–369

Maillet E (1905) Essais d'hydraulique fluviale. Herman, Paris

Manjoie A (1984) Karst superficiel dans la craie de la Hesbaye. In: Burger A, Dubertret L (eds) Hydrogeology of karstic terrains, vol 1. UNESCO, pp 70–71

Markova OL (1967) The influence of karst on the maximum flow on the rivers of the East European plains, vol 1. Proc Dubrovnik Symp Hydrol of Fractured Rocks, pp 387–401

Meyer FW (1984) Disposal of liquid wastes in cavernous dolostones beneath Southeastern Florida. In: Burger A, Dubertret L (eds) Hydrogeology of karst terrains, vol 1. UNESCO, pp 211–216

Mijatović B (1968) Metoda ispitivanja hidrodinamičkog režima kraških izdani pomoću analiza kriva pražnjenja i fluktuacije nivoa izdani u recesionim uslovima. (Method for hydrodynamic regime investigation using recession curve and water level fluctuation). Vesnik, Ser B, 8:43–81

Mijatović B (1984a) Captage par galerie dans la région de Herceg Novi (Yugoslaviae). In: Burger A, Dubertret L (eds) Hydrogeology of karstic terrains, vol 1. UNESCO, pp 126–129

Mijatović B (1984b) Captage par galerie dans un aquifère karstique de la côte Dalmate: Rimski bunar, Trogir (Yugoslaviae). In: Burger A, Dubertret L (eds) Hydrogeology of karstic terrains, vol 1. pp 152–155

Mikulec S, Bagarić I (1966) Izbor najpogodnije metode mjerenja kapaciteta ponora (Choice of the most suitable method for measuring sink capacity). Radovi i Saopštenja, Sarajevo 6:119–130

Mikulec S, Trumić A (1976) Engineering works in karst regions in Yugoslavia. Proc Karst hydrology and water resources. WRP: 443–488

Milanović P (1981) Karst hydrogeology. Water Res Publ p 434

Milanović P (1983) Some methods of hydrogeologic exploration and water regulation in the Dinaric karst with special reference to their application in Eastern Herzegovina. In: Hydrogeology of Dinaric karst. Geozavod, Beogr: 121–151

Milanović P (1986) Influence of the karst spring submergence of the karst aquifer regime. J Hydrol 84:141–156

Milanović P, Vučić M, Jokanović A (1985) Sanacija kaverne ispod tunelske cijevi. (Cavern sanation below tunnel pipe). Gradevinar 37, (9):371–380

Miličević M (1976) Influence of reservoir on changes in natural flooding of upstream karst plains. Proc Karst hydrology and water resources. WRP:387–408

Mimikou M, Kaemaki S (1985) Regionalization of flow duration characteristics. J Hydrol 82:77–91

Moore GK, Burchett CR, Bingham RH (1969) Limestone hydrology in upper Stones river basin, Central Tennessee. Tennessee Dept of Conservation, Div of Water Resources

Morozov VI, Lushchik AV (1984) Artificial recharge of karst water in the Flat Crimea. In: Burger A, Dubertret L (eds) Hydrogeology of karstic terrains, vol 1. UNESCO, pp 185–188

Motyka J, Wilk Z (1984) Hydraulic structure of karst-fissured Triassic rocks in the vicinity of Olkusz (Poland). Kras i Speleologia 14, (5):11–24

Newson M (1973) The hydrology of limestone caves. Cave Sci J Brit Spel Assoc 50:1–12

Novak D (1971) A contribution to the knowledge of physical and chemical properties of the groundwater in the Slovenian Karst. Krš Jugoslavije (Carsus Iugoslavie) 7/5:171–188

Paloc H, Mijatović B (1984) Captage et utilisation de l'eau des aquifères karstiques. In: Burger A, Dubertret L (eds) Hydrogeology of karstic terrains, vol 1. pp 101–112

Paloc H, Thiery D (1984) Essai de simulation du comportement hydrodynamique d'un karst par modèle déterministe. In: Burger A, Dubertret L (eds) Hydrogeology of karstic terrains, vol 1. UNESCO pp 80–83

References

Petrik M (1961) Measurements on submarine springs. Proc of 2nd Yug kongres of speleology, pp 49–57

Petrović J (1983) Kraške vode Crne Gore (Montenegro Karst Water). Posebna izdanja Univerziteta u Novom Sadu, p 111

Pitty AF, Halliwell RA, Ternan JL, Whittel PA, Cooper RG (1979) The range of water temperature fluctuations in the limestone waters of the Central and Southern Pennines. J Hydrol 41:157–159

Potie L (1973) Etudes et captage des résurgences d'eau douce sous-marines. 2° Colloque Int eaux souterraines. Palermo, pp 603–620

Ristić D (1976) Water regime of flooded karst poljes. Proc Karst hydrology and water resources. WRP: 301–316

Roglić S, Bonacci O, Margeta J, Žic K (1985) Primjena Stanford modela na slivovima u kršu (The application of the Stanford Watershad Model on the Catchments in Karst). Proc 7th Int Symp CAD/CAM, Zagreb, pp 171–176

Šegota T (1968) Morska razina u halocenu i mladem Würmu (Sea level in Halocene and Würm). Geografski Glasnik, Zagreb, 30:15–39

Sket B, Velkovrh F (1981) Postojna and Planina caves system as model for investigations of groundwater polution. Naše Jame 22:27–43

Sodemann PC, Tysinger JE (1967) Effects of forest cover upon hydrologic characteristics of a small watershed in the limestone region of East Tennessee, vol 1. Proc of the Dubrovnik Symp Hydrol of Fractured Rocks, pp 139–151

Soulios G (1984) Infiltration efficace dans le karst hellénique (Effective infiltration into Greek karst). J Hydrol 75:343–356

Soulios G (1985) Recherches sur l'unité des systèmes aquifères karstiques d'après des exemples du karst Hellénique. (Research on the unity of karstic aquifer systems after examples of Hellenic karst). J Hydrol 81:333–354

Srebrenović D (1970) Problemi velikih voda (High waters problems). Tehnička knjiga, Zagreb, p 277

Stepinac A (1979) Vodni kapacitet u stijenskim masama krša. (Water storage capacity of karst rocks). Gradevinar 31 (6):251–258

Stringfield VT, LeGrand HE (1969) Relation of sea water to fresh water in carbonate rocks in coastal areas with special reference to Florida, USA, and Cephalonia (Kephallinia), Greece. J Hydrol 9:387–404

Stringfield VT, LeGrand HE (1971) Effects of karst features on circulation of water in carbonate rocks in coastal areas. J Hydrol 14:139–157

Sweeting M (1973) Some results and applications of karst hydrology, part 2. Geogr J 139:280–328

Šegota T (1968) Morska razina u halocenu i mladem Würmu (Sea level in Halocene and Würm). Geografski Glasnik, Zagreb, 30:15–39

Torbarov K (1975) Estimation of permeability and effective porosity in karst on the basis of recession curve analysis. Proc Karst hydrology and water resources. WRP: 121–136

Turc L (1954) Le bilan d'eau des soles. Troisième journée de l'Hydraulique, Alger: 36–43

UNESCO (1984) Guide to the hydrology of carbonate rocks. Studies and reports in hydrology 41:347

Vecchioli J, Ehrlich GG, Godsy EM, Pascale CA (1984) Alterations in the chemistry of an industrial waste liquid injected into limestone near Pensacola, Florida. In: Burger A, Dubertret L (eds) Hydrogeology of karstic terrains, vol 1. UNESCO, pp 217–221

Vlahović V (1972) Zavisnost proticanja rijeke Donje Zete od nivoa kraške izdani i poplava u Nikšičkom polju (Relationship between the Donja Zeta River discharge groundwater levels and flood levels in the Nikšičko Polje). Krš Jugoslavije 8/2:17–41

Vlahović V (1983) Kraška akumulacija Slano (Slano Karst Reservoir). Crnogorska Akad Nauka i Umetnosti, p 246

Waagen L (1910) Karsthydrographie und Wasserversorgung in Istrien. Z Prakt Geol 18:229–239

White WB (1969) Conceptual models for carbonate aquifers. Ground Water 7(3):15–21

White WB (1977) Conceptual models for carbonate aquifers: revisited. In: Dilamarter RR, Csallany SC (eds) Hydrologic problems in karst regions. Western Kentucky Univ, pp 176–187

White EL, White WB (1983) Karst landforms and drainage basin evaluation in the Obey river basin, North-Central Tennessee, USA. J Hydrol 61:69–82

Williams PW (1983) The role of subcutaneous zone in karst hydrology. J Hydrol 61:45–67
Yevjevich V (1955) Metode za odredivanje približnog bilansa voda zatvorenih i plavljenih kraških polja. (Some methods for the polje in the karst water balance determination). Zbornik radova 3 Instituta J Černi, pp 7–13
Žibret Ž, Šimunić Z (1976) A rapid method for determining water budget of enclosed and flooded karst plains. Proc Karst hydrology and water resources. WRP: 319–340
Žibrik K, Lewicki F, Pičinin A (1976) Hydrologic investigation. In: Gospodarič R, Habič P (eds) Underground water tracing, Investigations in Slovenia 1972–1975. SAZU Ljubljana, pp 43–55
Zötl JG (1974) Karst hydrological investigation for the construction of the Diessbach reservoir (Austria). In: Burger A, Dubertret L (eds) Hydrogeology of karstic terrains, vol 1. UNESCO, pp 164–166

Geographical and Subject Index

Adriatic Sea 1, 2, 49, 50, 63, 64, 89, 149, 160, 162, 163
Alabama 9
Albania 2, 62
Alpine Valley 62
America 1, 9
Argostoli 109
Asia 14
Australia 63
Austria 2, 62, 151
Avignon 53
Ayios Georgios Spring 173

Bačina (Bačinska) Lakes 19, 162–165
Base flow 19, 20
Belgium 47
Biele Vodỳ 169, 171
Bileća (Bilečko) Polje 153
Bileća Reservoir 62, 152–156
Black Sea 63
Blue Lake 12
Boljkovac Spring 63
Brackish spring 97–102
Brévine Valley 40, 47
Bročanac 16
Bulaž Spring 9, 36, 53, 54, 66, 85–87
Buško Blato Polje (Reservoir) 25, 47, 112, 145, 157–159

California 63
Canada 14, 148
Čapljina 40, 42
Carlsbad Cave 35
Catchment 81–97
Cave 17, 21, 143
Cerkničko (Cerknica) Polje 152, 153
Cetina River 25, 39, 43, 47, 53, 54, 56, 61, 66, 90–94, 131, 147, 157, 158
Cetinje 106–108
Cetinjsko (Cetinje) Polje 65, 106, 108
Channel (passage) 17, 38, 71, 75, 76, 79
Cheddar Spring 19
Chile 63
Clarinbridge River 167
Coefficient of transmissivity 8
Conduit flow 19, 20, 76, 89

Coutagne equation 86
Crna Gora (Montenegro) 4, 104–106
Crnojevića River 65, 67
Cuba 14, 63
Czechoslovakia 169, 171

Dabarsko Polje 47, 154, 155
Dalmatia 4, 19, 147
Diessback Reservoir 151
Diffuse flow 9, 19, 76
Dimensions of fissures 38
Dinaric karst 1, 2, 9, 12, 14, 19, 25, 39, 49, 58, 61, 64, 78, 83, 84, 89, 90, 109, 111, 131, 144, 146, 147, 153, 160
Discharge curve 129
  for springs 67–75
  for ponors 109–115
Distance (fictitious) velocity 8
Doline (sinkhole) 12
Dover 168
Drogue's structural model 26–30
Dry valley 12
Drying up period 18, 131–135
Dubrovnik 136
Dumanly Spring 65

Effective infiltration 120
Effective porosity 6, 36–48
Ervenik 13
Europe 1
Evapotranspiration 87

Factor of karst 121
Fatničko Polje 46, 47, 112, 114, 154–156
Filtration coefficient 40
Flint Ridges Cave 11
Flooding 65, 106, 136, 154–156
Florida 63, 100, 168, 169
Flow duration curve 122–123
France 14, 16, 47, 53, 62, 63, 144, 172, 173

Gacka River 116
Glenfield Area 142
Gornja Dobra River 125–127
Gorski Kotar 4

Grab Spring 47, 158
Gračačko Polje 104, 112
Grančarevo Dam 152
Grassy Cove 14
Gravitational water 6
Great Britain 9, 19, 47, 148, 167
Great Tatra 171
Greece 2, 14, 63, 109, 121, 122, 133, 173
Groundwater hydrograph 87–89
Groundwater level 19, 22–24, 67–74, 110, 113, 116, 120, 168–171

Hardness 76, 77, 79, 80
Hawaii 63
Herceg Novi 173
Herzegovina 4, 39, 46, 62, 153
Hesbaye 47
Hutovo 12
Hydrograph 19, 75–81, 93
Hydrologic budget 136–140

Ionian Sea 109
Ireland 167
Israel 63
Istra (Istria) 36, 53, 66, 83
Italy 2, 14, 62, 63, 173

Jadro River (Spring) 89, 90, 92, 146, 147, 158, 159
Jama 14–16, 21, 23
Jamaica 14, 63
Jankovića Buk 47
Japan 63
Jelar Ponor 104, 105

Karren 10
Karstification depth (erosione base) 22, 36–48
Kaštela Bay 64
Kentucky 9, 17
Kephallenia Island 109
Klokun Spring 19, 20, 45
Konavosko Polje 136, 138, 160
Krčić River 23–25, 49, 50, 56–58, 70, 71, 116–118, 123, 127, 128, 132–134
Krčić Spring 58, 59, 61, 132
Krka River 13, 49, 50, 56, 118, 147
Krka Springs 55, 57, 58, 60, 70, 71, 147
Krupa 147
Krupac Polje 16, 17
Krupački Ponors 16

Lamalou Spring 47
Laminar flow 9, 11, 75
Lez Spring 169
Libanon 47, 63
Libya 63, 121, 133

Lika River 116
Livanjsko Polje 157
Ljubljanica River 52, 53, 152
Ljuta 147

Mammouth Cave 17
Manavgat River 67
Mangaphoue Area 142
Marasovine 13
Marocco 14
Matica Vrgorska River 162
Mediterranean Sea (Basin) 14, 63, 150, 173
Mendip Hills 10, 19, 47
Mexico 63
Miami 168
Miljacka 13
Montpellier 169
Mrzlek Springs 67

Nahanni 14
Neretva River 12, 47, 147
Neuchâtel 39
Nevesinjsko Polje 140
New Zeland 142
New York 63
Nikšičko Polje 47, 61, 71–74

Obod Estavelle (Cave) 46, 65
Olkusz 17, 38, 39
Ombla Spring 47, 84, 147
Ontario Area 1, 148
Orlina 16
Overland flow 18
Oymapinar Dam 67, 149

Palanka 13
Pantan Spring 89, 99, 173
Pennine Region 148
Pensacola 169
Permeability 7, 40, 41
Persian Gulf 63
Phreatic zone 21, 22, 79
Pierre Saint Martin 15
Piezometer 23–25, 42, 68–74, 75, 81, 112, 113, 154
Pivka Cave 143, 144
Pivka River 52
Planinska Cave 32–35, 52, 142–144
Planinsko Polje 52
Poland 17, 38, 39
Polje 5, 13–15, 25, 106, 136–140, 151–156, 159–163
Pološka Cave 16
Ponikve 114
Ponor (swallow hole) 17, 25, 61, 72, 103–115, 119, 130

Geographical and Subject Index

Popovo Polje 104
Port-Miou 172, 173
Postojna Cave 17, 33, 52, 142, 143
Prančevići Reservoir 25, 47, 158, 159
Primary (textural or intergranular) porosity 5
Pulse wave 76, 77
Pyrenees 16

Quick flow 19, 20

Radobolja Spring 169
Rastok Polje 58
Recession curve 43–45, 78, 80, 154
Red Lake 12
Retained water 6
Rivers in karst 116–135, 145–148
Rječina Spring 47
Roman Well 173
Rošca 71, 72
Ruda River (Spring) 47, 53, 55
Rumin Spring 158
Runoff coefficient 85–86, 120, 121

Saalach Valley 151
Salakovac 47
Sea ponor 109
Secondary porosity 5
Shaft flow 18
Šilovka 92, 93
Simultaneous discharge measurements 124–129
Široka ulica 16
Skadar Lake 62
Slana Cave 17
Slano Polje 16
Slovenia 4, 148, 152
Soča River 67
Somerset 19, 47
Sorgenti d'Aurisina Spring 173
Spain 14, 16, 63, 133
Specific retention 6
Specific yield 6
Split 30, 66, 89
Spring 36, 37, 45, 49–102, 145–148, 158, 159, 169–173
   cave s. 69
   estavelle 60–62, 111
   eyes 66
   fissure s. 66
   hidden s. 69
   intermittent (temporary) s. 58–60, 82
   permanent s. 53–58, 82
   rhythmic s. 60
   sublacustrine s. 62
   submarine (vrulje) s. 49, 51, 63–65, 149, 171–173

systems 67
Srinjine Quarry 30, 31
Standford watershed model 90–93
Storage capacity 6
Storage coefficient 6
Studenci Spring 54–56
Subcutaneous flow 18
Subcutaneous zone 28–35, 79
Surface hydrotechnical works 167, 168
Sušica River 61
Swallow capacity 109–115
Swiss Jura 39
Switzerland 47, 62
Syria 63

Tara River 55
Tennessee 14, 47, 123
Terrien 144
Theis equation 46, 48
Throughflow 18
Tolmin 16
Total porosity 6
Tracer 10, 50, 75
Trebišnjica River 12, 46, 47, 84, 152, 154
Trickle 33, 34
Trieste 173
Tučić Ponor 103, 104
Tunis 14
Tunnel 39, 41, 42, 159–166
Turbulent flow 9, 11, 17, 19, 71, 74, 76, 80
Turc equation 86
Turkey 65, 67, 77, 133, 149

Unica River 52
United Kingdom 10, 19
Ural 40
USA 11, 14, 17, 63, 98, 100, 123, 168, 169
Usage of spring water 169–171
USSR 40, 123, 124, 169

Vadose flow 18
   seepage 18
   zone 21, 79
Vaucluse Spring 53
Velocity of water 8–10, 42, 80
Vidović Bridge 146
Vinalić 91–93
Virginia 9
Vrgorsko (Vrgorac) Polje 58, 59, 112, 161–163
Vrtac Polje 104–106
Vukovića Spring 147

Wales 148
Water losses along rivers 124–135
Water storage 151–159
Water temperature 64, 80, 141–149

Yorkshire 9
Yugoslavia 1, 2, 4, 9, 14, 16, 30, 33, 34, 36, 39, 40, 43, 46, 47, 49–57, 59, 62–67, 70–74, 83–85, 89, 90, 93–95, 99, 103–109, 111, 112, 114, 116–118, 121, 127, 128, 131–136, 138–140, 144, 147, 151–153, 156–158, 160–165, 169, 173

Zadar 63
Zakučac 39, 41
Žegar 47
Zeta River 9, 47, 67, 71–74
Zrmanja River 13, 47, 93–97, 133–135, 147
Žrnovnica Bay 49, 50, 62, 64
Žrnovnica Spring 51, 65, 89, 90, 147